To Anna
hope you enjoyed
your time in Australia
& hope we see you again soon
love & best wishes 2012
Phil & Janice xxxx

THE BUSINESS OF NATURE JOHN GOULD AND AUSTRALIA

THE BUSINESS OF NATURE
JOHN GOULD
AND AUSTRALIA

ROSLYN RUSSELL

NATIONAL LIBRARY OF AUSTRALIA

Published by the National Library of Australia
Canberra ACT 2600

© National Library of Australia 2011

Books published by the National Library of Australia further the Library's objectives to interpret and highlight the Library's collections and to support the creative work of the nation's writers and researchers.

Every reasonable endeavour has been made to contact the copyright holders. Where this has not been possible, the copyright holders are invited to contact the publisher.

This book is copyright in all countries subscribing to the Berne Convention. Apart from any fair dealing for the purpose of research, criticism or review, as permitted under the *Copyright Act 1968*, no part may be reproduced by any process without written permission. Enquiries should be made to the publisher.

National Library of Australia Cataloguing-in-Publication entry

Author:	Russell, Roslyn.
Title:	The business of nature : John Gould and Australia / Roslyn Russell.
Edition:	1st ed.
ISBN:	9780642276995 (hbk.)
Subjects:	Gould, John, 1804–1881.
	Animals--Australia--Pictorial works.
	Animals in art.
	Natural history illustration--Australia.
Dewey Number:	591.994

Text:	Roslyn Russell
Development:	Susan Hall
Picture research:	Felicity Harmey
Editor:	Tina Mattei
Designer:	Andrew Rankine, atypica
Printer:	Australian Book Connection

Front cover image:
Henry Constantine Richter, lithographer (1821–1902) and Gabriel Bayfield, colourist (1781–1871), after Elizabeth Gould (1804–1841)
Calyptorhynchus macrorhynchus
(Great-billed Black Cockatoo)
in *The Birds of Australia*, vol. V, by John Gould
(London: John Gould, 1848)

CONTENTS

Acknowledgments … vi
Introduction: 1804 … 1
From Lyme Regis to London … 6
To the Printed Page … 13
Natural History Impresario … 20
Gould's & Darwin's Finches … 27
To Tasmania … 31
In the Field: Tasmania, New South Wales & South Australia … 37
Among the Menuras in New South Wales … 45
Experiencing Loss … 53
Victorian Entrepreneur among the Hummingbirds … 62
Epilogue: Gould's Legacy in Australia … 70
Portfolio
Australian Birds … 74
Australian Mammals … 140
Appendix 1: Gould's Major Published Works … 214
Appendix 2: Gould's Australian Chronology 1838–1840 … 216
Index … 218

ACKNOWLEDGMENTS

I would like to thank the National Library of Australia Publications Manager, Susan Hall, for asking me to write this book, which has reawakened my childhood interest in natural history, birds in particular. Tina Mattei was a meticulous editor of the manuscript, Felicity Harmey assisted in identifying the images from Gould's works reproduced in the book, and Andrew Rankine provided a most pleasing design.

The staff of the Petherick Reading Room and the Manuscripts Collection of the National Library of Australia, the Mitchell Library in the State Library of New South Wales, the Library of the Natural History Museum, London, and Mary Godwin, Curator and Head of the Philpot Museum, Lyme Regis, United Kingdom, have all assisted in providing source material and information for this book. Former Mitchell Librarian, Elizabeth Ellis, also pointed me in some productive directions. Ingrid Persaud read and commented on every chapter of this book as it was emailed to her in Barbados, and kept me focused on both the timetable and the audience. My husband, Michael Jones, as ever, kept everything going in the background as I worked on the book. To all of them I extend my gratitude and thanks.

This book is dedicated to the memory of my father, Reginald Brice, who passed away in August 2010.

Roslyn Russell

INTRODUCTION
1804

On Friday 14 September 1804, Jane Austen wrote to her sister Cassandra from the seaside watering place of Lyme Regis, in Dorset:

> We are quite settled in our Lodgings by this time, as you may suppose, & everything goes on in the usual order. The servants behave very well & make no difficulties ... Hitherto the weather has been just what we could wish; ... the continuance of the dry Season is very necessary to our comfort.

She wrote in a postscript, 'The Bathing was so delightful this morning ... that I believe I staid in rather too long'.[1] On the same mild, early autumn day, but in very different social circumstances to those the future novelist enjoyed, John Gould was born in Lyme Regis to a gardener and his wife. That same year, on the other side of the world, the settlement that would become known as Hobart was established. In 1838, Gould and his partner in his work and in his life, Elizabeth, landed in Hobart, their first port of call in Australia. It was the beginning of a landmark scientific and artistic collaboration that would lead to the creation of two of the most beautiful and important early scientific publications documenting the unique fauna of Australia: *The Birds of Australia* and *The Mammals of Australia*.

Today, the town of Lyme Regis is a bustling place in the tourist season, particularly attracting two niche markets—literary and geological pilgrims. Lovers of literature come in their thousands to walk on the Cobb, the stone-built harbor wall that stretches its long arm into the sea, and recreate for themselves the moment in Austen's novel, *Persuasion*, of Louisa Musgrove's fall from the steep stairs to the ground; or stroll along the seaside promenade, where the waves of the English Channel suck up the pebbles and toss them back again in a sibilant rhythm. Fossil hunters fossick around the base of the cliffs in search of ammonites and other finds, following in the footsteps of Mary Anning, the fossil collector whose discoveries of fine specimens of ichthyosaurs and plesiosaurs between 1810 and 1840 gave to this part of Britain the title of 'Jurassic Coast'. Austen's and Anning's names are proudly displayed in the Philpot Museum as notable people with connections to Lyme Regis. They are joined by John Fowles, author of *The French Lieutenant's Woman*, the film version of which brought the town and the Cobb to the world's movie screens, and Thomas Coram, who established the Foundling

J. Walker, after J. Nixon
The Bay of Lyme Regis, Dorset 1796
engraving on paper
Courtesy Mary Evans Picture Library, London

Hospital, in London's Coram Fields.[2] Lyme Regis also commemorates John Gould as a celebrated person with a local connection. The Philpot Museum has some Gould lithographs on display and, in 2009, a large exhibition was held there that 'raised his profile considerably in Lyme'.[3]

At the time that Gould was born, the scientific community of Britain was undergoing significant cultural shifts. From being dominated by wealthy dilettante amateurs, often of aristocratic status, the community was passing into the hands of middle-class careerists, who increasingly looked to the various scientific disciplines to provide a basis for their professional lives. Sir Joseph Banks, President of the Royal Society, who had defined the role of the gentleman amateur scientist in Britain throughout the latter part of the eighteenth century, would loom over its scientific establishment for another 16 years until his death in 1820, by which time the new men of science were already building their careers. Broadly based scientific societies, such as the Royal Society and the Linnaean Society of London, were joined by organisations catering for specific branches of natural history. The eighteenth-century interest in botany was now extended to zoology, leading to a focus on separate branches of the subject, in particular, for the purpose of this book, the study of birds—ornithology.

M. & N. Hanhart, London, after Thomas Herbert Maguire (1821–1895)
Portrait of John Gould, Ornithologist 1849
lithograph on paper; 29.3 x 24.0 cm
Pictures Collection, nla.pic-an9547887

Animal and plant specimens continued to pour in from collectors all around Britain's expanding empire and their preservation for display and study became an urgent consideration. More effective taxidermy materials using arsenic had become widely available by 1830 and led first to better preserved specimens and then to an enhanced capacity for precise identification and classification. Artistic advances, such as the adoption of lithography for book illustration, also provided expanding opportunities for those motivated to take commercial advantage of them, assisted by new developments in printing technology. At the same time, natural history enjoyed an unprecedented boom in popularity among the general public, described as 'a deeper, more constant enthusiasm for natural history in all its branches which gathered strength from year to year from the 1820s through to the 1860s, and which touched every section of

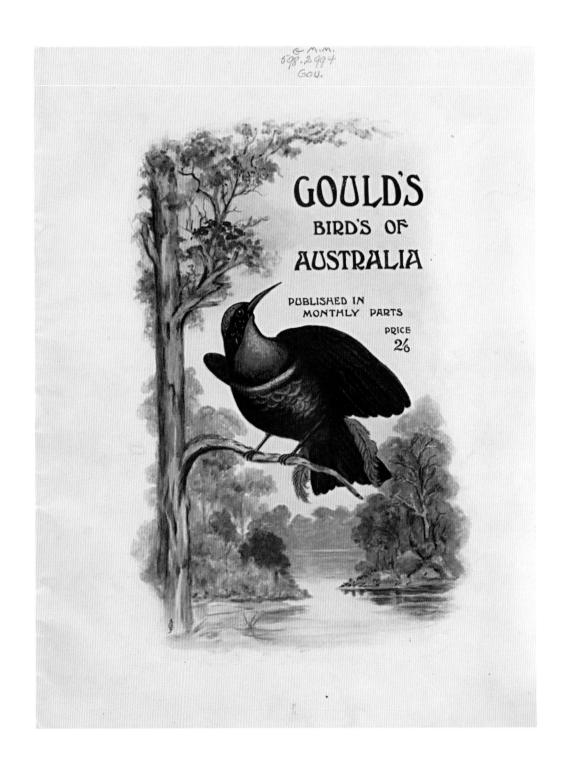

Front cover of *Gould's Birds of Australia: A Fac-simile Reproduction in a Reduced Form* **reproduced by G.J. Broinowski**

(Sydney: Simmons-Bloxham Printing, 1910)

society'.[4] The conjunction of all these circumstances provided the context in which the young Gould was able to gain a foothold and then attain eminence in the world of nineteenth-century British ornithology and publishing.

The boy born in Lyme Regis in humble circumstances would embark on a career of profound significance to ornithology and science generally, and would achieve enduring renown in the Antipodes as the 'father of Australian ornithology'. Gould's identification of finches brought back on HMS *Beagle* from the Galapagos Islands would provide Charles Darwin with one of the most important keys to unlocking the mystery of the origin of species. Gould would travel to the far-flung colony of Van Diemen's Land (Tasmania), where he and his wife, Elizabeth, would stay at Government House, in Hobart, as guests of Sir John Franklin and his redoubtable wife, Lady Franklin. Gould's son, Franklin, would be born there and be given the Governor's name. Gould and his collectors would travel around mainland Australia, several of the latter paying with their lives for their commitment to natural history collecting and exploration. His artists would depict in exquisite lithographs, accompanied by Gould's commentary, the birds and mammals of the island continent of Australia and of other parts of the world. Gould would sell these wonderful productions to subscribers with all the acumen of the Victorian businessman that he was. He would introduce into Britain and Europe the best-loved of all cage birds, a little Australian parakeet, the budgerigar. In Australia, Gould's name and legacy would be perpetuated in the Gould League, which celebrated its centenary in 2009. Through the League's work, Gould would become known to generations of Australians as the 'Bird Man'.

Over the course of his life, Gould's work as an 'illustrators' impresario'[5] and a skilled describer of bird and mammal species lifted him far above the social setting into which he was born. The story of his remarkable life—his practical skills, his driving energy and shrewd business judgement, his conspicuous talent for determining and describing the characteristics of birds and animals, his travels to locate, classify and illustrate new species, and his interactions with those with whom he worked and did business—has been told many times. This book owes a great debt of gratitude to those scholars who have compiled and edited his correspondence, and studied and written about Gould, and their work is acknowledged in the citations that accompany each chapter. This book is intended to provide, in its turn, a brief sketch of the life of the man whose classic volumes on the Australian fauna, *The Birds of Australia* and *The Mammals of Australia*, and other material are treasures of the National Library of Australia.

ENDNOTES

1. Deirdre Le Faye (ed.), *Jane Austen's Letters*, third edition, Oxford University Press, Oxford, 1997: 93, 95.

2. 'Mary Anning Recollected' in Judith Pascoe, *The Hummingbird Cabinet: A Rare and Curious History of Romantic Collectors*, Cornell University Press, Ithaca, New York, 2006: 140–168. In addition to her focus on Mary Anning, Pascoe also mentions other famous past residents of Lyme Regis, including the 'hummingbird obsessive John Gould': 147.

3. Email from Mary Godwin, Curator and Head of the Philpot Museum, Lyme Regis, 14 November 2009.

4. Lynn Barber, *The Heyday of Natural History, 1820–1870*, Jonathan Cape, London, 1980: 13.

5. *ibid.*: 95.

CHAPTER 1
FROM LYME REGIS TO LONDON

John Gould's was not the genteel middle-class environment of Jane Austen's or the prosperous and intellectual background of his contemporary and colleague, Charles Darwin. Gould was the eldest son of a gardener, who was said to have been particularly adept at raising cucumbers.[1] Gould senior was one of that vast army in Britain whose work provided not only food from the kitchen gardens of landed estates and suburban villas, but also maintained the ordered domestic cultural landscapes which have become familiar as the settings for televised costume dramas dealing with the lives of the gentry of early nineteenth-century England.

After working as a gardener in Lyme Regis, possibly for the Dowager Lady Poulett, whose house in the town boasted 'grand and sophisticated glass-houses',[2] Gould's father by his mid-twenties had received a better offer at a more extensive property, the Stoke Hill estate, near Guildford, Surrey. The advantages of this move were many, including, after some years, a higher status for the elder John Gould. The Stoke Hill estate contained a variety of landscape features—parklands, a lake, ornamental trees and shrubs, a kitchen garden and a conservatory—thus supplying a diversity of tasks for a gardener keen to master the range of skills associated with his occupation. It also provided what Gould junior's great-great-granddaughter would later call 'happy memories of the wild commons and heaths of the Surrey countryside', by making these natural areas accessible to the young boy.[3]

Reflecting on his childhood later in life, Gould cast a romantic light over his first remembered encounter with the natural world, crediting his infant experience as the motivating force for his life's endeavours:

> How well do I remember the day when my father lifted me by the arms to look into the nest of the Hedgesparrow in a shrub in our garden! This first sight of its beautiful verditer-blue eggs has never been forgotten; from that moment I became enamoured with nature and her charming attributes; it was then I received an impulse which has not only never lost its

Benjamin Fawcett (1808–1893),
after Alexander Francis Lydon (1836–1917)
Ripley Castle, Yorkshire 1879 (detail)
from *A Series of Picturesque Views of Seats of the Noblemen and Gentlemen of Great Britain and Ireland*, vol. 1, edited by Reverend F.O. Morris
(London: William MacKenzie, 1879)
Baxter print on paper
Courtesy Mary Evans Picture Library

influence, but which has gone on acquiring new force through a long life.[4]

Gould's reverence for nature's 'charming attributes', however, did not make him overly sensitive about the ways in which its productions, including his beloved birds, were collected for the purpose of advancing scientific knowledge. Despite his words of regret about his boyhood activity of despoiling birds' nests for their eggs, to be blown and strung up proudly on his cottage walls, only to be 'vigorously destroyed in our games before the termination of the year, in order to ward off the ill-luck otherwise supposed to ensue',[5] Gould was a willing participant in the exploitation of the natural world throughout his life—in the cause of science, to be sure, but also the enrichment of himself. The death of countless birds and animals was, however regrettable, a necessary concomitant of the study of natural history in the centuries before the invention of binoculars, high-speed cameras and video recorders. Nevertheless, one of Gould's much-admired predecessors in the art of nature illustration in Britain, Thomas Bewick, swore off killing animals after his sensitive heart was wrung by the death of a hare in a hunt and of a bullfinch at which he had thrown stones.[6] No such squeamishness troubled Gould, although he deplored the needless slaughter of birds and animals, writing in *The Birds of Great Britain* that: 'throughout the work I have been a champion for our poor persecuted birds and defended them as well as I could in words on account of the great amount of good they effect'.[7] Gould was by no means the only celebrated ornithologist to combine the talents of a hunter with a love of birds: his great contemporary, John James Audubon, has been described as an 'avid hunter' and 'a lover and observer of birds and nature'.[8]

The nature of Gould's formal education is unknown but it could only have lasted until his early teens; his appalling handwriting caused dismay to his correspondents, in an age when the better educated tended to write in a neat copperplate hand. As an adult he deployed a somewhat pompous and characteristically Victorian prose style, replete with classical allusions. It is clear, though, from Gould's subsequent career and from the number of scientific papers he produced, not to mention the extensive commentary in his natural history publications, that he had a keen intelligence. He has been described as a 'genius'.[9] He had acute powers of observation and discernment, and an extraordinary memory for the essential features of a bird or mammal and the environment in which it lived and functioned.

In 1818, two things happened to Gould that helped to shape his adult life. Firstly, his father was offered a job in the Royal Gardens at Windsor Castle and the young John, his years of formal education deemed to be at an end, was apprenticed as a gardener. According to the *Gardeners' Chronicle* in 1881: 'Between the ages of fourteen and twenty he spent most of his time under the care of the late J.T. Aiton, at the Royal Gardens, Windsor, where he acquired a taste for botany and horticulture'.[10] This is not strictly true—Gould may have retained some interest in his former occupation but, later in life, he admitted: 'Botany is not my forte'.[11] Ornithology, however, *was* Gould's forte. His friend, Richard Bowdler Sharpe, recorded: 'He had begun to study birds in earnest, and he made the acquaintance of many British species for the first time in the royal domain'.[12]

The second circumstance to have an effect on Gould's life was contingent upon the first. As an apprentice gardener at Windsor, he was taught a very practical skill, one closely aligned with his interest in the natural world and birds in particular—taxidermy. Learning to stuff birds led Gould to the career in ornithology that he would pursue with unremitting zeal for the rest of his life. It did not take him long to capitalise on this newly acquired skill, as he is reported as having begun his entrepreneurial career in his first year at Windsor, catching and stuffing birds and selling them to Eton scholars for the Boys' Museum at the famous school located within view of the castle. He was adept at taxidermy from the beginning. In an age where prepared specimens usually degraded quickly, Gould's lasted: the British Museum still has a pair of magpies which he stuffed in the first year of his apprenticeship.[13]

Between mid-September 1823 and late February 1825, Gould was sent to work on another stately property, Ripley Castle, in Yorkshire. As well as learning how to grow plants under glass in the estate's greenhouses, Gould also continued his taxidermy and study of birds.[14] By the age of 21, however, he had decided that gardening would not be his lifetime profession, given the low pay, hard work and few prospects of achieving wealth. He decided that it was time to go into business for himself in London.

Using his skills as a taxidermist, Gould quickly obtained high-profile commissions. Within a few months, he became the first taxidermist to be patronised by British royalty: he mounted a 'Thick Knee'd Bustard' (now known by a rather more attractive name, stone curlew or *Burhinus oedicnemus*) for no less a personage than King George IV.[15] Institutional and individual clients also brought their specimens to the young taxidermist in Broad Street, Golden Square, Soho, for stuffing and mounting for their museums. A listing of Gould's clients reads as a 'Who's Who' of the natural history establishment of early nineteenth-century England: the British Museum, the United Services Museum, the East India Company, the Royal College of Surgeons, the Earl of Derby and Thomas Campbell Eyton.[16]

Gould's skill in taxidermy and the new contacts he was making in the scientific world, where his sound advice on what specimens to obtain made him sought after, paid off. A new horizon was about to open up for the ambitious and highly motivated young man.

On 29 April 1826, the Zoological Society of London was formed by explorer, collector and founder of Singapore Sir Stamford Raffles, scientist Sir Humphry Davy and, among others, zoologist and Irish politician Nicholas Aylward Vigors, who was the Society's secretary until 1833. Frustrated by the Linnaean Society's perceived emphasis on botany (unsurprising in view of its name), the group of men who founded the Zoological Society wanted to focus on the animal world. To this end, they established collections of both living and dead animal species for the purpose of scientific study, the former in Regents Park and the latter in Bruton Street, London.

Either in 1827 or in early 1828, after a competitive examination, Gould was appointed to the position of Curator and Preserver at the Museum of the Zoological Society of London.[17] To be appointed to the position at such a young age 'would suggest that Gould was already demonstrating he was much more than merely a bird-stuffer'.[18] Indeed, he showed that he was more than a mere 'bird-stuffer' when he 'staked his first claim to fame by stuffing a giraffe'.[19] The giraffe, a gift from Mehemet Ali, Pasha of Egypt, was the special pride of its owner, the King. When the animal died in October 1829, after two years in the foreign climate of England, Gould and another colleague at the Zoological Society, Tomkins, were given the task of stuffing it in preparation for its donation to the Society. The success of this delicate operation sealed Gould's reputation.[20] At the end of Gould's life in 1881, over 50 years since he went to work at the Society, Bowdler Sharpe reported of the young curator that:

> My friend Mr. Gerrard remembers him in these early days as a man of singular energy, with a good knowledge of the art of

mounting animals, and indeed some of the best taxidermists in England were working under Gould at that time—such men as Baker, Gilbert, and others.[21]

Gould had established himself in a strong position in the scientific community of London and beyond.

In a few years, Gould had moved from a position on the lower rungs of the social ladder of Regency Britain to mixing as a professional with men of far higher social status. His next move even more firmly cemented his position in the ranks of the middle classes. In 1829, he married Elizabeth Coxen, a governess described by Bowdler Sharpe in 1893 as 'the daughter of a Kentish gentleman'. The last word conveyed a wealth of social meaning to a nineteenth-century reader.[22]

Elizabeth Coxen was born in another English seaside resort, Ramsgate, on 18 July 1804, the third surviving child of Nicholas and Elizabeth Coxen (née Tomkins), a middle-class couple with naval and military connections. Elizabeth was the only girl in the family to attain maturity. Two of her brothers, Stephen and Charles, had both already migrated to Australia by the time she and Gould met and married, and both would play parts in the story of John Gould. In her early twenties, Elizabeth was condemned to the fate of many an educated middle-class young woman and was working as a governess teaching Latin, French and music (and maybe art also) to the nine-year-old daughter of William Rothery, Chief of the Office of the King's Proctor in Doctors' Commons, in 18 James Street, Buckingham Palace.[23] The lonely young woman wrote to her mother: 'I feel I shall get very melancholy here', and told her of her longing to know someone who could 'enter one's feelings'. Although her young charge was 'a perfect child', Elizabeth often found life 'wretchedly dull'.[24]

Their great-great-granddaughter, Maureen Lambourne, has speculated that Elizabeth Coxen and John Gould first met in the temporary aviary of the Zoological Society in Bruton Street.[25] Isabella Tree, in her biography of John Gould, advances a more prosaic explanation of how these future life partners and artistic collaborators met and formed a relationship. A visit by Gould to a near relation of Elizabeth's, a Mr Coxen, who worked as a birdstuffer in Broad Street, Golden Square, close to where Gould lived, was most likely, considers Tree, to have been the occasion on which the two probably first met. Whatever the circumstances, Tree writes, the connection worked brilliantly to Gould's advantage: 'Elizabeth was determined, intelligent, educated, practical, obedient, and she possessed the one attribute he most desperately lacked—she could draw'.[26]

'Women were trained to draw well: it was one of the few things they practised and studied consistently from infancy through to adulthood', writes Lynn Barber in *The Heyday of Natural History*.[27] Drawing and painting in watercolour were two of the 'accomplishments' expected of middle- and upper-class women in Regency and Victorian Britain, along with a range of other skills designed to fill endless hours in a period where those with even quite modest claims to gentility were expected to apply themselves to reading and a range of domestic crafts.[28] Attaining these skills also helped a young woman to appear to advantage on the marriage market and to assist her husband in his social role thereafter (along with bearing his children with monotonous and often life-threatening regularity). For educated women who did not marry but whose families could not afford to support them through long years of spinsterhood, becoming a governess was one of the few employment options available. It was then the governess' task to impart to her charge or charges the accomplishments that she herself had learned, particularly if the household in which she served could not support a separate drawing or music teacher.

Elizabeth was rescued by marriage from the isolation, often tedious work and indeterminate social status of being a governess. After the couple wed on 5 January 1829, at St James Church, Piccadilly, Elizabeth found that, despite her married status and frequent childbearing, she *was* expected to work, albeit at an occupation that did not violate the code of gentility—drawing birds on lithographic stone to her husband's directions, a task to which she brought her considerable skills and dedication. Instead of enduring soul-destroying boredom, she travelled to Europe and to Australia, and met with a wide range of people and situations. She and her husband appeared to be 'soul mates' as they worked together in their business and raised a growing family. Years after her death, she received the ultimate accolade for a Victorian wife from Bowdler Sharpe: 'to this lady is due much of the ultimate success of her husband's career, for she was as accomplished an artist as she was one of the best of wives'.[29]

Mrs. John Gould. From a painting in the possession of her great-grandson, Dr. Alan Edelsten, England.

Unknown artist
Mrs John Gould
Private Collection

ENDNOTES

1. Ann Datta, *John Gould in Australia: Letters and Drawings*, Miegunyah Press, Carlton, Victoria, 1997: 23, records that John Gould senior's obituary noted 'he was remembered as a skilled horticulturalist, and, in the growth of the cucumber, excelled by none'.
2. Isabella Tree, *The Bird Man: The Extraordinary Story of John Gould*, first published 1991, this edition Ebury Press, London, 2004: 9, cites the Philpot Museum, Lyme Regis, as the source for Gould senior's possible employment by the Dowager Lady Poulett.
3. Maureen Lambourne, *John Gould: Bird Man*, Osberton Productions, Milton Keynes, 1987: 19.
4. John Gould, *The Birds of Great Britain*, quoted in Datta, *op. cit.*: 29.
5. *ibid.*: 30.
6. Keith Thomas, *Man and the Natural World: A History of the Modern Sensibility*, Pantheon Books, New York, 1983: 177, 280.
7. John Gould, *The Birds of Great Britain*, quoted in Gordon C. Sauer, *John Gould, the Bird Man: A Chronology and Bibliography*, Henry Sotheran Limited, London, 1982: xvi.
8. 'John James Audubon 1785–1851: The American Woodsman: Our Namesake and Inspiration', The Audubon Society website, www.audubon.org/nas/jja.html, viewed 4 January 2010.
9. Sauer, *op. cit.*: xvii.
10. *Gardeners' Chronicle*, 1881: 217, quoted in Ray Desmond, *Kew: The History of the Royal Botanic Gardens*, The Harvill Press with The Royal Botanic Gardens Kew, London, 1995: 431. John Townsend Aiton was the son of William Townsend Aiton, the head gardener at Kew and the author of the *Hortus Kewensis*, 1789. The quotation is probably from an obituary of Gould, as it appeared in the year of his death, 1881.
11. Tree, *op. cit.*: 12.
12. Richard Bowdler Sharpe, *An Analytical Index to the Works of the Late John Gould, F.R.S., with a Biographical Memoir and Portrait*, Henry Sotheran, London, 1893: x.
13. Tree, *op. cit.*: 13; Bowdler Sharpe, *op. cit.*: x, reported that at Windsor 'he had already established a reputation for skilful taxidermy'.
14. Lambourne, *op. cit.*: 22; Datta, *op. cit.*: 28.
15. Datta, *ibid.*: 30.
16. *ibid.*: 34.
17. Datta gives April 1828 as the date when Gould took up the position; other sources, such as Lambourne, *op. cit.*: 23, say 1827.
18. Datta, *op. cit.*: 42.
19. Tree, *op. cit.*: 1. Tree opens her lively biography of John Gould with a chapter on his treatment of the royal giraffe, and its provenance and significance.
20. Datta, *ibid.*: 30.
21. Bowdler Sharpe, *op. cit:* x.
22. *ibid.*: xi.
23. Datta, *op. cit.*: 64–65.
24. Lambourne, *op. cit.*: 26.
25. *ibid.*: 25.
26. Tree, *op. cit.*: 32.
27. Lynn Barber, *The Heyday of Natural History, 1820–1870*, Jonathan Cape, London, 1980: 126.
28. *ibid.*: 16–20. Lynn Barber writes of the capacity of natural history to fill in the time of an affluent Victorian family with 'rational amusement': 'there was nothing the middle classes need so badly as something to do'.
29. Bowdler Sharpe, *op. cit.*: xi.

CHAPTER 2
TO THE PRINTED PAGE

For most of the 1830s, John Gould combined his private work as a taxidermist with his role as Curator and Preserver at the Zoological Society of London, until he resigned in early 1838 before travelling to Australia. Gould's career throughout the 1830s shows that he was able to juggle the competing demands of employment and entrepreneurship.[1] Working at the Society gave him the opportunity to identify people of talent who would later assist him in his work, including fellow taxidermist John Gilbert. It also plugged him into a network of natural history enthusiasts who formed the basis for the subscription list for his future publications and assisted him in other ways.

That Gould was acceptable to the scientific community had been demonstrated by his appointment to the Zoological Society. Although the position was still lowly, it was a giant step upwards for the former gardener. Further evidence of this came in 1830, when Gould's first scientific paper was published in *Zoological Journal*, the journal of the Zoological Club of the Linnaean Society, which preceded the Zoological Society, but closed in 1829. This paper was the first of over 300 that he would publish in his lifetime. When the Zoological Society began publishing its own *Proceedings* in 1831, Gould lost no time in furthering his scientific reputation, submitting at least five papers in the 1832 issue and, from then on, appearing regularly as a contributor.[2]

It would, however, be wrong to imagine that it was Gould's natural talents and growing scientific reputation that brought him so rapidly to the Zoological Society's notice as a potential contributor of scientific papers. The Society did pride itself on being open to those outside the ranks of the aristocracy and regarded itself as being different from the rest of the club-like scientific societies due to 'the proportion of men of science and sound principles which it contains'.[3] It was, however, the support of the Society's Honorary Secretary, Nicholas Aylward Vigors, that enabled Gould to leap across the social and professional divide and take his place as a published author in the field of ornithology.

Vigors' befriending of the young John Gould provided advantages beyond the publication of Gould's scientific papers. Described as 'a fanatical zoologist with a special interest in birds',[4] Vigors was a Member of Parliament for the Irish constituency of County Carlow. He and Gould began to establish a distinctive focus on birds within the Zoological Society, culminating in Gould's appointment to the position of superintendent of a separate ornithological department within the Society in 1833.[5] Vigors' friendship and support also gave Gould the entrée to the society of other eminent ornithologists, who would also become 'intimate friends', 'invaluable allies' and the first subscribers to his illustrated publications on birds. Among them were Prideaux John Selby and Sir William Jardine, eminent naturalists who had already published books on birds; the latter became Gould's 'first serious patron'.[6]

It was the combination of employment at the Zoological Society, a talent for roughly sketching the appearance and attitude of a bird,[7] his desire to collect bird specimens, and the friendship and support of his eminent colleagues that first led Gould into publishing illustrated works of ornithology. An interesting collection of Himalaya Mountains bird skins that contained some ornithological 'novelties' came into Gould's possession in around 1830. Vigors undertook the scientific descriptions of the birds for the Zoological Society and then Gould, according to Richard Bowdler Sharpe, had a brainwave: 'the idea struck Gould that an illustrated work might be published, with figures of these interesting birds. Vigors would write the letterpress and Gould himself would be responsible for the illustrations'.[8]

Gould recognised that he did not yet have the scientific standing to identify, name and classify the birds with the rigour required by systematic ornithology so that their description would be acknowledged by the scientific community. However, the combination of quality illustrations with text written by the respected Vigors, would, Gould decided, produce a publication that could credibly enter the growing market for works of natural history.[9] He intended to achieve this by means of the new reproduction technique of lithography. The work itself was to be named *A Century of Birds from the Himalaya Mountains*, as there were to be a hundred birds depicted in the book.

M. & N. Hanhart, London, after Thomas Herbert Maguire (1821–1895)
Sir William Jardine 1851 (original 1849)
lithograph on paper
© National Portrait Gallery, London

But who would make the fine drawings of these one hundred birds on the lithographic stones? Herein lay the problem: Gould was by no means a polished artist who could produce the fine artistic renderings demanded for publication of works on natural history. Bowdler Sharpe was exquisitely tactful when describing the strengths and weaknesses of Gould's artistic technique:

> No one has excelled Mr. Gould in his appreciation of bird-life. He was in every way an ornithologist and knew and loved the birds. He was always able to sketch, somewhat roughly perhaps, the positions in which the

birds were to be drawn upon the plates, and no one could have a better 'eye' for specific differences.[10]

Other writers were less tactful: Sacheverell Sitwell commented that 'it must be understood that Gould was but a poor artist himself';[11] and Isabella Tree writes that some of his drawings were 'childlike in their simplicity and awkwardness'.[12] Gould's initial sketches of birds on which the polished lithographic plates were based nevertheless have their defenders. His great-great-granddaughter, Maureen Lambourne, claims:

> Gould's quick drawings, in pencil, ink and broad colour wash, have directness, substantiality and a bludgeoning strength of character sometimes lost in the finished plates. They are often heavily annotated with colour notes and instructions, as if Gould, aware of his own artistic shortcomings, was anxious that all the finest points of the bird should be recorded.[13]

Australian naturalist Allan McEvey saw 'the directive influence of Gould's sketches as a genuine contribution to the art of British natural history' and 'that in the rough vitality of Gould's original sketches there was a quality not surpassed in the finishing techniques of the artists of his workshop, and that consequently a creative and inspiring role is to be attributed to them'.[14] Tree, who considers that Gould had preserved his 'inadequacy as a draughtsman' as a 'well-kept secret' and perpetuated the myth of himself as an artist, admits: 'No matter how displeasing they may seem, Gould's drawings were an integral part of the production; despite their amateurishness, they convey a significant impression of form and substance'.[15]

One thing is certain: Gould was well aware of his artistic limitations when it came to producing the fine illustrations required for publication using lithography and he recognised that he would have to recruit artists to undertake this work. He did not have to look far. Bowdler Sharpe captured the moment in his memoir of Gould:

> With his unfailing instinct he rightly estimated his wife's artistic powers, and he broached the subject to her. 'But who will do the plates on stone?' she asked; for lithography was not at that time an everyday matter as it is now. 'Who?' replied her husband. 'Why *you*, of course!' This was the story as told me by my old friend forty years after the event. Anyhow, aided by the sketches of her husband, Mrs. Gould did draw all the plates of the 'Century', and, with the exception of Bewick's woodcuts and the beautiful illustrations of Swainson, no such clever drawings of birds had been seen in this country.[16]

While lithographic drawing was regarded as a ladylike occupation, bending over a block of limestone cannot have been easy for Elizabeth: during her depiction of the Himalaya Mountain birds she gave birth to a son, John Henry, a few days after the first plates of the publication were sent to subscribers in December 1830. Gould wrote to Jardine the day before the birth, hoping for his 'approbation and support' of the work, and noting, somewhat anxiously: 'It is Mrs. Goulds very first attempt at stone drawing which I hope you will take into consideration'.[17] Her artistry, aside from some comments about inaccuracy in representing the toes of pheasants, met with the approval of Gould's colleagues in the ornithological community, such as Selby and Jardine. The former commented that the plates 'are very well done upon the whole … I like them as well as Audubon's'.[18] This was high praise indeed, especially for a neophyte such as Elizabeth was when it came to drawing on lithographic stones.

Vigors, the author of the text, was so appreciative of Elizabeth's work on *A Century of Birds from the Himalaya Mountains* (for which she drew 102

right
**Elizabeth Gould
(1804–1841)**

Muscicapa melanops

from *A Century of Birds from the Himalaya Mountains* by John Gould (London: Printed by C. Hullmandel, 1832)

far right
**Elizabeth Gould
(1804–1841) and
Edward Lear
(1812–1888)**

Muscipeta princeps

from *A Century of Birds from the Himalaya Mountains* by John Gould (London: Printed by C. Hullmandel, 1832)

right
**Elizabeth Gould
(1804–1841)**

Myophonus horsfieldii

from *A Century of Birds from the Himalaya Mountains* by John Gould (London: Printed by C. Hullmandel, 1832)

far right
**Elizabeth Gould
(1804–1841) and
Edward Lear
(1812–1888)**

Fringilla rodopepla and *Fringilla rodochroa*

from *A Century of Birds from the Himalaya Mountains* by John Gould (London: Printed by C. Hullmandel, 1832)

left
Elizabeth Gould
(1804–1841)

Picus shorii

from *A Century of Birds from the Himalaya Mountains* by John Gould (London: Printed by C. Hullmandel, 1832)

far left
Elizabeth Gould
(1804–1841)

Bucco grandis

from *A Century of Birds from the Himalaya Mountains* by John Gould (London: Printed by C. Hullmandel, 1832)

left
Elizabeth Gould
(1804–1841)

Tragopan satyrus

from *A Century of Birds from the Himalaya Mountains* by John Gould (London: Printed by C. Hullmandel, 1832)

far left
Elizabeth Gould
(1804–1841)

Lophophorus impevanus

from *A Century of Birds from the Himalaya Mountains* by John Gould (London: Printed by C. Hullmandel, 1832) figures of birds on 80 plates,[19] based on drawings by John of around 90 specimens in his own collection and the remainder from other collections) that he bestowed her name on a sun-bird, *Cinnyris gouldiae* or *Nectarinia gouldiae*.[20] Gould, however, was not so impressed by Vigors' inability to deliver the text until 18 months after all the plates had been completed, owing to his political commitments. From now on, Gould would undertake all the writing himself. 'Never again', writes Tree, 'would Gould allow himself to be compromised by someone else's priorities'.[21]

ENDNOTES

1. Conflict-of-interest ethics that apply to today's museum professionals would have nipped some of Gould's business activities in the bud. The Code of Ethics of the International Council on Museums is very clear on this issue: *Dealing in Natural or Cultural Heritage* (paragraph 8.14) reads: 'Members of the museum profession should not participate directly or indirectly in dealing (buying or selling for profit), in the natural or cultural heritage'.
2. Ann Datta, *John Gould in Australia: Letters and Drawings*, Miegunyah Press, Carlton, Victoria, 1997: 51.
3. Datta, *op. cit.*: 53; Isabella Tree, *The Bird Man: The Extraordinary Story of John Gould*, first published 1991, this edition Ebury Press, London, 2004: 20. The Society's degree of inclusivity even extended to women with the appropriate scientific background.
4. Tree, *ibid.*: 19.
5. *ibid.*: 23.
6. *ibid.*: 27–28.
7. Today's ornithologists use the term 'jizz' as shorthand for 'the appearance and demeanour of a bird'. Although this word does not appear in any of the literature on Gould, his talent was for capturing just this quality and being able to describe it. Libby Robin, Robert Heinsohn and Leo Joseph (eds), *Boom & Bust: Bird Stories for a Dry Country*, CSIRO Publishing, Collingwood, Victoria, 2009: x.
8. Richard Bowdler Sharpe, *An Analytical Index to the Works of the Late John Gould, F.R.S., with a Biographical Memoir and Portrait*, Henry Sotheran, London, 1893: xi–xii.
9. Datta, *op. cit.*: 70–71.
10. Bowdler Sharpe, *op. cit*: xxiii.
11. Sacheverell Sitwell, *Tropical Birds from Plates by John Gould*, B.T. Batsford Limited, London, 1948: 12.
12. Tree, *op. cit.*: 32.
13. Maureen Lambourne, *John Gould: Bird Man*, Osberton Productions, London, 1987: 67–68.
14. J.T. Burke, foreword to Allan McEvey, *John Gould's Contribution to British Art: A Note on Its Authenticity*, Sydney University Press for the Australian Academy of the Humanities, *Art Monograph* 2, 1973.
15. Tree, *op. cit.*: 33.
16. Bowdler Sharpe, *op. cit.*: xii.
17. Quoted in Tree, *op. cit.*: 36.
18. Quoted in *ibid.*: 37. John James Audubon was the doyen of bird illustration in the United States, and was also celebrated in Britain and Europe for his vivid illustrations of birds, all life-size.
19. The two extras were chicks of species already depicted, so there were 100 birds represented in the book. Datta, *op. cit.*: 71.
20. Tree, *op. cit.*: 37–38.
21. *ibid.*: 38.

Elizabeth Gould (1804–1841)
Cinnyris Gouldiae
from *A Century of Birds from the Himalaya Mountains* by John Gould
(London: Printed by C. Hullmandel, 1832)

CINNYRIS GOULDIÆ.

CHAPTER 3
NATURAL HISTORY IMPRESARIO

A *Century of Birds from the Himalaya Mountains* signalled that John Gould was a force to be reckoned with in ornithological illustration in Britain. To achieve this, he had joined forces, albeit temporarily, with another rising star in bird illustration in Britain, Edward Lear.

Lear, better known now as the author of nonsense verse, in particular the ever-popular poem 'The Owl and the Pussycat', was the polar opposite of the bluff, energetic Gould. He came at the tail end of the family of a London stockbroker whose fortunes on the Exchange had fallen into ruin. The twentieth child in a family of 21, the young Edward was brought up by his sister, Ann. She and another sister, Sarah, took charge of their brother's basic education, read him poetry and taught him to draw. Epileptic and depressive, Lear was a highly talented artist, particularly of natural history subjects.

Forced to earn his living in his mid-teens, Lear turned to his talent for illustration. He worked for naturalists, such as Prideaux John Selby and Sir William Jardine, and, like Gould, gained the support of Nicholas Vigors. Encouraged by these mentors, at the age of only 18, Lear embarked on an ambitious project—to illustrate all the species of the parrot family, the *Psittacidae*. Whenever

Unknown artist
Edward Lear 1830s
silhouette on paper
© National Portrait Gallery, London

possible, he used his preferred method of capturing the birds' appearances by drawing them from life in the Parrot House at the Zoological Gardens and preparing his drawings for printing by using the as-yet-uncommon technique of lithography. Lear was a highly accomplished illustrator but, not unexpectedly, given his youth and temperament, a poor businessman. The first two folios he published in November 1830 brought him instant recognition as an ornithological artist and he received the accolade of nomination to associate status in the revered Linnaean Society of London.

Producing fine works of natural history illustration, however, required the assistance of a number of other people and trades, and the coordinating skills to keep the process on track. Lear, creative and talented artistically, found that extracting from subscribers the money necessary to publish the next set of plates was so difficult that he was forced to find paying work elsewhere. John Gould, in need of another artist, especially one of Lear's calibre, stepped into the picture. Indeed, Gould's choice to adopt lithography as his reproduction technique for *A Century of Birds from the Himalaya Mountains* had been inspired by Lear's example. In 1831, Lear began to work with Gould on the latter's next ambitious ornithological publishing venture, *The Birds of Europe*, travelling with John and Elizabeth throughout Europe, visiting zoos and museums across the continent, and observing and sketching birds in the wild. He worked alongside Elizabeth and to him goes the credit for teaching her the finer points of lithography. Lear also took over the work during one of the infrequent occasions when Elizabeth could not continue with the inexorable process of drawing birds on stone for her husband's publications, as she had just lost a child.[1]

Lear's work in *The Birds of Europe,* for which he drew 62 of the 448 plates, is justly celebrated. One in particular, that of the eagle owl (*Bubo maximus*), is a powerful image of a magnificent bird that confronts the viewer directly, at a time when most bird illustrators showed them in profile. The moon is reflected in one of its huge golden eyes, and the intricate detail of the feathers around the head, chest, wings and lapping its feet is delicately rendered. Lear's fondness for birds that could be portrayed as characters, such as the owl that found its way into his most famous nonsense rhyme of all, is evident in his illustrations for *The Birds of Europe*.

Lear, however, found Gould to be a relentless taskmaster and, when the opportunity arose in 1832 to escape to Lord Stanley's estate at Knowsley Hall, to draw the animals in His Lordship's menagerie, he did so. He was, nonetheless, committed to finishing his quota for *The Birds of Europe* for Gould and, in 1833, he agreed to work on another Gould ornithological publication, contributing ten plates to *Monograph of the* Ramphistidae, *or Family of Toucans*, published between 1833 and 1835.

But Lear's artistic interests were evolving in a different direction. He had travelled to Ireland in 1835 and developed an enthusiasm for landscape painting. A year later, his eyesight began to fail, ruling out the close work required for natural history illustration. By 1837, he had begun to travel back and forth from England to the continent, a pattern he continued, with excursions further afield, for many years. Lear also began to publish in those areas in which his name is now best known: nonsense verse and landscape studies. His only work of natural history after he finished his association with Gould was *Gleanings from the Menagerie and Aviary at Knowsley Hall*, published in 1846. Lear tried to maintain his relationship with Gould by letter but was always disappointed by the perfunctory scrawled responses he received in answer to his own effusive missives. Thirty years after he had worked with Gould, Lear delivered a verdict on him that clearly stemmed from disappointment at the failure of their relationship to survive time and distance, but which has contributed to an often-stated view of Gould as a somewhat unsympathetic character: 'A more singularly offensive mannered man than G. hardly can be: but the queer fellow means well, tho's more of an Egotist than can be described'.² And after Gould's death, Lear fired a parting shot that further damned his ex-artistic collaborator as 'a harsh and violent man', and summed up his view of his working life with Gould, who was, he declared:

> ever the same persevering hardworking toiler in his own (ornithological) line,—but ever as unfeeling for those about him. In the earliest phase of his bird drawing he owed everything to his excellent wife, & to myself,—without whose help in drawing he had done nothing.³

This is one view of Gould that has endured but other voices tell a different story. It is difficult to imagine that Gould could have achieved what he did had he been as unattractive a personality as Lear's strictures suggest. Managerial expertise only goes so far. All the complex effort that was required to maintain the uninterrupted flow of plates and text to subscribers meant not only smoothly coordinated activity but also loyalty from Gould's assistants. He was clearly a man who was able to inspire others with his vision, starting with his wife. They were a devoted couple whose joint activity in ornithological illustration created a strong bond in addition to their roles as spouses and parents.⁴ They were also complementary personalities: 'Elizabeth Gould's quiet but sympathetic manner, her sensitivity and her cultured background must have been in sharp contrast to her husband's assertive and bustling ways', writes the couple's great-great-granddaughter.⁵

The next person to commit to John Gould's enterprises was a 20-year-old clerk, Edwin Charles Prince, who came to work for Gould soon after he and Elizabeth had moved from Bruton Street to 20 Broad Street, Soho; Prince remained in the position for the rest of his life until he died at age 64 in 1874. Described by Richard Bowdler Sharpe as 'a most able and conscientious man',⁶ he was responsible for all of

Edward Lear (1812–1888)
Bubo maximus (Eagle Owl)
in *The Birds of Europe* by John Gould (London: John Gould, 1837)
Image courtesy of the Division of Rare and Manuscript Collections, Cornell University Libraries

Gould's business affairs, encompassing not only the publication of illustrated works of natural history but also a flourishing taxidermy practice. One of Gould's later illustrators, the celebrated natural history artist, Joseph Wolf, summarised Prince's contribution to Gould's success: 'without this timely help and the ceaseless labour of a long-suffering slave, one Prince, (called by courtesy a "Secretary"), the result might have been different'.⁷

Elizabeth Gould (1804–1841)
Trogon macrouna (Large-tailed Trogon) 1838
hand-coloured lithograph on paper
Courtesy Natural History Museum, London

Prince himself did not subscribe to Wolf's patronising view of his role in Gould's business. In a letter to Gould on 5 February 1840, in which he thanked Gould for acknowledging his work, Prince showed himself to be more than willing to exert himself on his employer's behalf: 'Believe me it will always be my highest gratification to merit the good opinion of everyone but none more than yourself, and the more confidence you repose in me the more strenuous will be my efforts to prove myself worthy of it'.[8] Also, any perusal of correspondence from Prince to Gould, such as the many letters he wrote to his absent employer when the latter was travelling, demonstrates that he in no way addressed Gould in a servile tone.

What was the process by which a Gould publication developed, from its conception to its delivery to subscribers? It is useful to trace the steps between the initial idea for a publication and its successful execution, to understand the contributions to the finished product of John Gould, Elizabeth and other artists, Prince and a range of trades and businesses.

First of all, Gould identified an ornithological subject: it could be a particular genus or the birds of a specific region. Next, a prospectus describing the projected publication was sent by Prince to potential subscribers. Gould's contacts through the Zoological Society of London were invaluable in this regard and he had no compunction about presenting his publications to royalty and then heading up the list of subscribers with their names. The publication of fine illustrations carried out by Gould was expensive and sales on an open market were unpredictable, so subscriptions were necessary to finance a publication. It did not go ahead until these financial requirements had been met.

Gould himself would write the text and draw a rough sketch of the bird (or animal, in the case of *The Mammals of Australia*), mostly from a stuffed specimen but with his vision informed by his observation of the living bird or animal where this was possible. The sketch would capture its essential physical characteristics and attitude, often accompanied by notes to indicate his intentions even more strongly to his artists. While their contribution was crucial to the quality of the finished plate, 'still his was always the moving spirit in designing the plates', wrote Bowdler Sharpe.[9]

The sketch then went to Elizabeth, Lear and later to other artists, for a refined version to be drawn with a special greasy crayon as a reversed image onto a slab of

John Gould
(1804–1881)
Euphema splendida
Gould 1846

pencil and crayon on paper;
53.0 x 37.5 cm
Pictures Collection,
nla.pic-an9994496

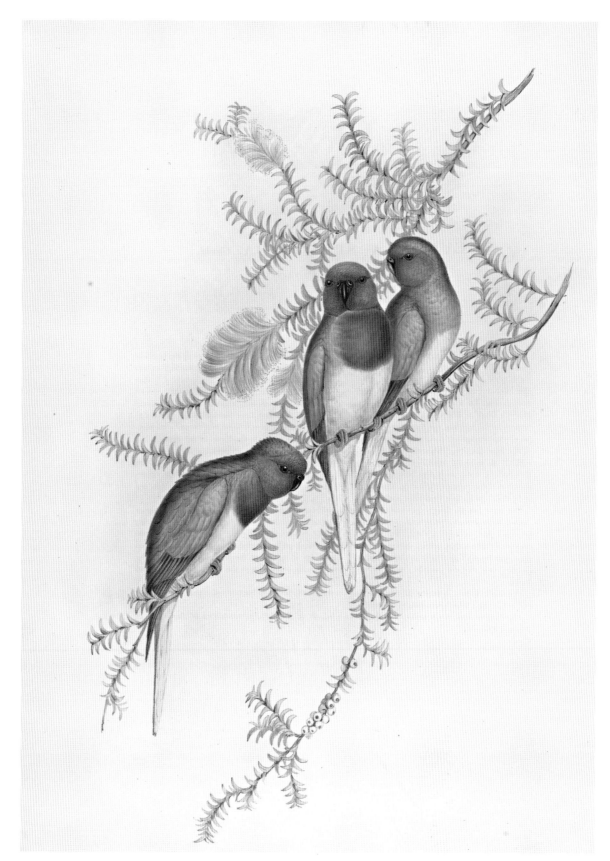

Elizabeth Gould (1804–1841), after John Gould (1804–1881)

Euphema splendida (Splendid Grass Parrakeet)

in The Birds of Australia, vol. III, by John Gould (London: John Gould, 1848)

John Gould (1804–1881)
Moorhens with Young among Waterlily and Reeds
pencil and gouache on board; 39.0 x 57.0 cm
Pictures Collection, nla.pic-vn3068089

smooth limestone. The stone went to the lithographic printer, Charles Joseph Hullmandel, who printed copies in tones of black. These copies were then taken to colourists, along with a hand-painted example that they were to follow. Gabriel Bayfield and his family were Gould's colourists for many years. The text (or letterpress as it was known) written by Gould was copied out by Prince in his meticulous handwriting and taken by him to the printer.

Once the illustrations had returned from the colourist and the text from the printer, Prince compiled them in the correct order and sent them to the subscribers. They were designed not to be bound into a volume until all the parts had been issued. Gould then sent out precise bookbinding instructions, a necessary condition of the publication of scientific work, where correct ordering of plates and text by systematic arrangement was crucial.

By the end of the 1830s, Gould had put into place the structures that underpinned his lifetime achievement as the outstanding British natural history impresario of the nineteenth century. His dual business interests—natural history illustration and publication, and taxidermy—were supported by talented and industrious people. His ornithological vision and his ability to delegate different aspects of the work to those with the ability to carry it out ensured the smooth running of a complex business operation. His single-minded devotion to his work was the linchpin holding the whole edifice together. The public appetite for natural history provided the niche market for the productions that flowed in a ceaseless stream from Gould's home-based business.

So began a lifelong career of ornithological, scientific and artistic entrepreneurship, and a reputation that has survived to the present day. The gardener's son would interact in his professional capacity with aristocrats and members of Britain's governing class, a measure of the social mobility that could result from a combination of talent and industry. Expanding his artistic and scientific career in the years just before Queen Victoria's reign commenced in 1837, Gould was a quintessential Victorian entrepreneur, ever ready to take advantage of the opportunities that the expansionist age put in his way.

ENDNOTES

1. Isabella Tree, *The Bird Man: The Extraordinary Story of John Gould*, first published 1991, this edition Ebury Press, London, 2004, devotes a chapter, 'Lear's Years of Misery': 43–63, to the contribution made by Lear to Gould's initial success.
2. *ibid.*: 63.
3. Quoted in *ibid.*: 43–44.
4. Ann Datta, *John Gould in Australia: Letters and Drawings*, Miegunyah Press, Carlton, Victoria, 1997: 65.
5. Maureen Lambourne, *John Gould: Bird Man*, Osberton Productions, London, 1987: 26.
6. Richard Bowdler Sharpe, *An Analytical Index to the Works of the Late John Gould, F.R.S., with a Biographical Memoir and Portrait*, Henry Sotheran, London, 1893: xvi.
7. Quoted in Maureen Lambourne and Christine E. Jackson, 'Mr Prince: John Gould's Invaluable Secretary', *Naturae*, no. 4, 1993, Centre for Bibliographical and Textual Studies, Monash University, Clayton, Victoria: 17.
8. *ibid.*: 17–18.
9. Bowdler Sharpe, *op. cit.*: xii.

CHAPTER 4
GOULD'S & DARWIN'S FINCHES

Throughout the time that John Gould was placing his combined ornithological publication and taxidermy business on a firm foundation, he was also still employed by the Zoological Society of London. His work there contributed, however unwittingly, to nineteenth-century science's greatest advancement in the world of ideas: the publication in 1859 of Charles Darwin's *On the Origin of Species*.

This event, with its momentous consequences, was over two decades away when a young naturalist, Charles Darwin, returned in October 1836 from the five-year voyage of HMS *Beagle* that had taken him, among other places, to South America, the Galapagos Islands and Australia. In all these places, Darwin had observed and collected animals and plants, the further study of which would lead him to the conclusion that species had evolved through a process of natural selection and had not been created in their fixed forms for all time.

Darwin had to find a repository for the many bird and animal skins which he had collected on the *Beagle* voyage. At first, he considered depositing them with the British Museum but its natural history collections were known to be in a poor state, so he offered them to the Zoological Society's

George Richmond (1809–1896)
The Young Charles Darwin 1840
© English Heritage Photo Library
Courtesy Darwin Heirlooms Trust

museum. There, the 450 birds that Darwin had collected came before the eyes of Gould, whose task it was, as the Society's ornithological specialist, to describe their characteristics, determine the species to which they belonged and to name them according to the Linnaean system.

Two types of birds in Darwin's collection caught Gould's attention. First of all were the finches that Darwin had brought back from the Galapagos Islands. Gould lost no time in describing and classifying these small and, in appearance, rather nondescript birds. He had received Darwin's birds on 4 January 1837 and, a mere six days later, presented the first batch of his scientific descriptions to the Zoological Society. Speed was of the essence if your ambition was to have the honour of describing and bestowing a name on a species, and Gould was ever eager for this recognition.[1]

The 12 species of finches from the Galapagos Islands described by Gould to the assembled fellows of the Zoological Society can probably be regarded as among the most famous birds in history. While it is clear that Darwin recognised their biological significance after the *Beagle* voyage and used this as a key argument to support his theory of evolution, it was Gould, the expert ornithologist, who determined their separate status as unique species, principally on the basis of the different shapes of their beaks.[2]

The second type of bird was a small ostrich from South America that had been all but eaten by Darwin's shipmates before the young naturalist had snatched away the remains, preserving the head, neck, legs and wings, the skin and many of the larger feathers. Gould had reconstructed the fragments and identified the bird as a distinct species, on which he conferred Darwin's name, *Rhea darwinii*.

Twenty-two years later, when Darwin published *On the Origin of Species*, he cited Gould's

Elizabeth Gould (1804–1841)
Geospiza scandens (Common Cactus Finch)
in *Birds* by John Gould, part 3 of *The Zoology of the Voyage of H.M.S. Beagle: Under the Command of Captain FitzRoy, R.N., during the Years 1832 to 1836*, edited and superintended by Charles Darwin (London: Smith, Elder and Co., 1840–1843)

identification of the birds of the Galapagos Islands as distinct species as a key piece of evidence for his theory of the transmutation of species, better known to us now as 'evolution by means of natural selection':

> The most striking and important fact for us in regard to the inhabitants of islands, is their affinity to those of the nearest mainland, without being actually the same species. Numerous instances could be given of this fact. I will give only one, that of the Galapagos Archipelago, situated under the equator, between 500 and 600 miles from the shores of South America. Here almost every product of the land and water bears the unmistakeable

In early January 1837, Gould accepted the task of writing the 'Birds' volume of *The Zoology of the Voyage of H.M.S. Beagle*, in collaboration with Darwin. He was hampered, though, by the small number of specimens that Darwin had collected and also by the very small amount of geographical and observational data he had gathered, particularly on the Galapagos finches. Darwin had, in a number of cases, neglected to note the locations from which the birds had been obtained, so Gould had to seek out more bird skins and information from other members of the *Beagle*'s company, including Captain Robert FitzRoy and Darwin's servant, Syms Covington. This slowed down Gould's progress on the text. Nevertheless, one component of the task was completed by the Gould family business. Elizabeth Gould was responsible for drawing on lithographic stone the birds that Darwin had collected during the voyage of the *Beagle*. In gratitude for her fine rendering of the 50 plates in the 'Birds' volume, Darwin wrote in the introduction: 'The accompanying illustrations, which are fifty in number, were taken from sketches made by Mr. Gould himself, and executed on stone by Mrs. Gould, with that admirable success, which has attended all her works'.[4] A year after John Gould had agreed to work with Darwin on the 'Birds' volume, though, he had begun to make other plans. Their realisation meant that the volume would have to be completed by another naturalist, this being carried out by the British Museum's ornithologist, George Robert Gray.[5]

An interest in Australia had been germinating in Gould's mind even before Darwin had returned with his accounts of the fauna of the southern continent. Elizabeth's two brothers, Charles and Stephen Coxen, were already living there, in the Hunter River district of New South Wales, at a property called 'Yarrundi'. They had been sending considerable numbers of bird skins to Gould, who was beginning to feel that he had some familiarity

Elizabeth Gould (1804–1841)

Camarhychus parvulus (Small Tree Finch)

in *Birds* by John Gould, part 3 of *The Zoology of the Voyage of H.M.S. Beagle: Under the Command of Captain FitzRoy, R.N., during the Years 1832 to 1836*, edited and superintended by Charles Darwin (London: Smith, Elder and Co., 1840–1843)

stamp of the American continent. There are twenty-six land birds, and twenty-five of those are ranked by Mr Gould as distinct species, supposed to have been created here; yet the close affinity of most of these birds to American species in every character, in their habits, gestures, and tones of voice, was manifest.[3]

The largest group of birds among those that Darwin referred to in this passage was the Galapagos finches.

Elizabeth Gould (1804–1841)
Rhea darwinii (Darwin's Rhea)
in *Birds* by John Gould, part 3 of *The Zoology of the Voyage of H.M.S. Beagle: Under the Command of Captain Fitzroy, R.N., during the Years 1832 to 1836*, edited and superintended by Charles Darwin (London: Smith, Elder and Co., 1840–1843)

with the birds of the Australian continent. Gould was sufficiently confident that he could prepare a publication on the subject that, in late 1836, he asked his friend and patron, Sir William Jardine: 'would not a work on the Birds of Australia be interesting?' Some plates of the Australian species had already been prepared, wrote Gould, and he was thinking 'of making it my next illustrative work'.[6]

He was also thinking of a more ambitious project, one that would mean an upheaval for himself, Elizabeth and their family, and also would effectively bring to an end his appointment at the Zoological Society. His letter to Jardine sets out exactly what he would do in less than two years from the time he penned it: 'I have even some serious intention of visiting the colony for two years for the purpose of making observations etc.'[7] The encounter with Darwin's specimens reinforced his intentions even further. By early 1838, his plans for the voyage to Australia were all but made and the step was about to be taken that led ultimately to John and Elizabeth Gould becoming known as 'the father and mother of Australian ornithology'. To speed them on their way and to thank them for their efforts, Darwin gave the departing couple two presents: a silver compass and a dram bottle.[8]

ENDNOTES

1 Ann Datta, *John Gould in Australia: Letters and Drawings*, Miegunyah Press, Carlton, Victoria: 84. Both Datta and Isabella Tree devote a chapter to this story in their biographies of Gould.

2 *ibid.* Darwin described another finch species at a subsequent meeting of the Zoological Society. Modern ornithology has reduced Darwin's original 13 Galapagos finch species to nine.

3 Charles Darwin, *On the Origin of Species by Means of Natural Selection, or, The Preservation of Favoured Races in the Struggle for Life*, first published by John Murray, 1859, this Penguin Classics edition, London, 1985: 385.

4 *The Zoology of the Voyage of H.M.S. Beagle*, introduction by Charles Darwin, quoted in Datta, *op. cit.*: 88.

5 *ibid.*

6 *ibid.* Gould did begin a publication and issued some plates, which he then recalled and cancelled so that he could prepare a more authoritative work on Australian birds. These 'cancelled parts' are now major collectors' items and fetch high prices.

7 *ibid.*

8 *ibid.*; Tree called her chapter on the Darwin connection 'A Dram Bottle from Darwin'.

CHAPTER 5
TO TASMANIA

John Gould had asked his wife to make her greatest sacrifice to his work and art. She agreed to leave behind their three beloved younger children and to travel with him to Australia to prepare the work on the birds of that distant country. The youngest of the Goulds' children, Louisa, was under six months old and sickly. She and her two-year-old sister, Eliza (also called Lizzie), would be cared for by Elizabeth's mother and cousin, Mrs Coxen and Mrs Mitchell, at the Goulds' family home in Broad Street. Four-year-old Charley was sent away to school. The usually resilient Elizabeth temporarily faltered under the distress and anxiety of leaving her small children and she suddenly became ill. For a short time, the whole enterprise was in doubt.[1]

Nevertheless, on 16 May 1838, John and Elizabeth Gould set sail for Australia on a small vessel, the *Parsee,* which carried a maximum of only 12 cabin passengers. Travelling with them was their eldest son, John Henry, who was seven and a half years old. Their 14-year-old nephew, Henry Coxen, whose father had died in 1825, came with them to join his uncles, Charles and Stephen Coxen, in New South Wales. Gould's colleague from the Zoological Society of London, John Gilbert, was brought along as collector. He was an inspired choice, as Gould alone could not hope to cover all the areas of Australia necessary for a comprehensive treatment of its birds, and Gilbert was a highly competent and trusted assistant. A man-servant, James Benstead, was also included in the party, as was Mary Watson,[2] who was to act as a companion to Elizabeth, a duty she had performed before for ladies travelling to India. The Goulds planned to be away from England, home and family for two years.[3]

The ever-energetic John Gould did not regard the sea voyage to Australia as a holiday. In a letter sent to his patron, Sir William Jardine, not long before he sailed, Gould set out the arrangements he had made, both for his wife's comfort and so that he could continue his work while aboard ship:

> I have a Stern Cabin eleven feet by twelve principally for the comfort of Mrs. G. and one adjoining of a smaller size. I have had the former fitted up with Drawers for birds and books in hope of work going along and taking a quantity of birds with me as well as left over plates to cut up for box books etc.[4]

Gould was not disappointed in his expectation that he could use his shipboard time productively.

Louis Auguste de Sainson (1801–1887), after Felix Achille Saint-Aulaire (1801–c.1889)
Vue de la rade de Hobart-Town, Ile Van-Diemende 1833
plate 51 in *Voyage de la corvette l'Astrolabe* (Paris: J. Tastu, 1830–1835)
hand-coloured lithograph on paper; 22.5 x 33.2 cm
Pictures Collection, nla.pic-an8368168

By mid-June he reported in a letter that 'the whole of our party are quite well' and, although they had 'experienced a few disasters and events' of 'little consequence' in bad weather, this 'had the effect of making us all good sailors … The past is quite forgotten and we are all cheerful and happy'. Gould was in a perpetual state of delight at the range of birds and other sea life that he could observe and capture, and had found so much to amuse him 'that I have scarcely read a book and have written but little'.[5] Time passed happily for Gould and he managed to obtain specimens of close to 30 species of sea birds caught on the open ocean during the voyage.

The Goulds' first destination in Australia was Tasmania. Word of the arrival of the celebrated ornithologist and his wife reached the small colony when the *Parsee* docked on 19 September 1838 after a four-month voyage. *Hobart Town Courier* announced their presence there with *éclat*: 'Mr and Mrs Gould are now in the Colony to which they have come at great expense and sacrifice of comfort'. The newspaper explained that the drawings for the projected work on the birds of Australia would be taken 'from living specimens'.[6] It is rather ironic that, in the first of 11 surviving letters that Elizabeth wrote to her mother and cousin during her absence, she described her husband as employing his usual method of securing bird specimens. He had,

Louis Haghe (1806–1885), after Joseph Mathias Negelen (1792–1870)
Portrait of Sir John Franklin 1849
lithograph on paper; 34.6 x 25.4 cm
Pictures Collection, nla.pic-an9579248

William Paul Dowling (c.1824–1877), after Thomas Bock (1793–1855)
Lady Franklin between 1868 and 1874
photographic copy of an original sketch (c.1840); 9.0 x 5.7 cm
Courtesy W.L. Crowther Library, Tasmanian Archive and Heritage Office

according to his wife: 'already shown himself a great enemy to the feathered tribe, having shot a great many beautiful birds and robbed various others of their nests and eggs'.[7]

In the same letter, Elizabeth reassured herself as much as her mother that: 'As for the dear children, I feel perfectly satisfied that they will not miss me, for I know well that you and my dear Mrs. Mitchell will do all for them that I could wish done'. Then the devoted mother got the better of the stoic, as she asked: 'I hope my dear little Louisa is not suffering from her teeth. How does my Charley like school and does Lizzy look as rosy as before her illness?'[8] Elizabeth probably knew by the time she wrote this letter that she was pregnant again and that her next child would be born in the Australian colonies.

The town to which the Goulds had come—known as Hobart, Hobart Town or Hobarton—was the main settlement of a colony that had existed for precisely the same span of years as their own lives. Thirty-four years since its foundation in the only landmass in the Australian colonies that consistently reminded

visiting English people of their homeland, but which still carried the Dutch name of Van Diemen's Land, the settlement of Hobart had yet to grow from a town into a city.

The Goulds' colleague, Charles Darwin, had visited Hobart just over two years before, on the homeward leg of the voyage of HMS *Beagle*, and his diary records his comparison of its urban area with that of Sydney. Hobart's 'streets are fine & broad; but the houses rather scattered: the shops appeared good'; and around Sullivan's Cove 'there are some fine warehouses; & on one side a small Fort'. When Darwin compared the town with Sydney, he was 'chiefly struck by the comparative fewness of the large houses, either built or building'. He attributed

Charles Edward Stanley (1819–1849)
Government House, Hobarton, V.D.L. c.1849
watercolour on paper; 26.3 x 36.4 cm
Pictures Collection, nla.pic-an3083383

this to the probability that fewer people, including ex-convict entrepreneurs, were amassing the vast fortunes that had produced lavish dwellings in Sydney. There was, Darwin wrote, a negative side to the Hobart situation: 'The growth however of small houses has been most abundant; & the vast number of little red brick dwellings, scattered on the hill behind the town, sadly destroys its picturesque appearance'.[9]

The Goulds moved into lodgings in Davey Street, in the centre of Hobart. They had come with letters of introduction to Sir John and Lady Franklin, the colony's Governor and his wife. At first, Lady Franklin confided to her husband in a letter, she had not managed to persuade the 'unassuming' Elizabeth to call on her, 'though various hints have been given her'. Sir John's *aide-de-camp*, Henry Elliott, 'our shrewd observer of manners', thought that 'Mr. Gould seems fully conscious of his importance as a lion', Lady Franklin wrote. She took a less judgmental view of John Gould than Elliott did: 'I do not see why a bird fancier should come all the way to the Antipodes in search of his peculiar fame, and not think the better of himself for it'.[10]

While the confident John Gould was socially at ease with the Franklins, Elizabeth was not so accustomed to mixing with people of such a high status, although her manners were more than adequate to the situation. There was, possibly, another reason for her initial shyness when it came to socialising in the Governor's circle: Sir John was a celebrity and his wife was a woman of considerable force of character and intellect. Elizabeth could have felt overpowered by Lady Franklin's personality.

Sir John's fame derived from his exploits as an Arctic explorer. With the goal of finding the much-sought-for North-West Passage, he had led two expeditions to the American Arctic in the 1820s, during the first of which he had faced starvation, and was afterwards famous for having been forced to boil up and eat his leather boots in order to survive. His second Arctic expedition earned for him the gold medal of the Geographical Society of Paris but his failure, yet again, to locate the North-West Passage rankled.[11] As a peacetime naval officer—he had served in the battles of Copenhagen and Trafalgar in the Napoleonic Wars and, before the second of these actions, had sailed in Matthew Flinders' *Investigator* as a midshipman—he was looking for employment, and was offered and accepted the governorship of Tasmania, to succeed Colonel Sir George Arthur.

Lady Franklin, energetic, idealistic and full of plans for the social and cultural advancement of the colony her husband had come to govern,[12] soon proposed an expedition on which both the Goulds were invited. Her goal was to reach Port Davey, in southwest Tasmania[13] and, if possible, to continue on to the more remote and wilder region of Macquarie Harbour. Elizabeth's pregnancy probably made her reconsider another sea voyage, even around the coast of Tasmania. She opted to stay in Hobart, while her husband joined the Governor's wife and others on the government schooner, the *Tamar*. By this time, she had become friends with Lady Franklin. Elizabeth wrote to her mother in December 1838 with the news that:

> We have been frequent visitors at Government House and have been staying also at their cottage, New Norfolk. I have also been invited to take up my abode at Government House during Lady F's. absence at Macquarie. Sir John, his niece, and some other ladies staying there, Lady F. thought it would be less lonely for me being in lodgings solitary.[14]

In the same letter she poured out questions to her mother about the little children she had left behind in England. Anyone who read her letter would be in no doubt that her heart was aching for them:

> My dear little Louisa too is just at a critical age, teething in all probability. I did not forget the darling's birthday. Bless their dear little faces. How I love to recall their looks to my mind's eye. Does Charles feel happy at school? I hope Mrs. Berryman will not press him too much with his books. I do not think he is at present of an age to make it important. My dear little Eliza, too, is she Mrs. Mitchell's pet or yours or

both? Do not, my dear Mother, *quite spoil* the dear children.[15]

Elizabeth kept herself busy while John was away: 'drawing as many native plants as I can, I mean branches of trees, some of which are very pretty'.[16] She wanted to capture on paper the environments in which she and her husband would depict the birds and animals discovered by him in Australia. Elizabeth also had a keen eye for the characteristics of human society in the Australian colonies. She noticed how expensive it was to live there and how difficult it was to obtain good servants. She commented on the motive that had brought many settlers to the colonies: 'The fact is that most persons come here with a determination to get money and return to England as soon as they can … This has been a famous place for money making—and I think money spending'.[17]

ENDNOTES

1. Isabella Tree, *The Bird Man: The Extraordinary Story of John Gould*, first published 1991, this edition Ebury Press, London, 2004: 88.
2. James Benstead and Mary Watson were later to return to Australia in the household of Sir George Grey and to marry each other.
3. Alec H. Chisholm, *The Story of Elizabeth Gould*, Hawthorne Press, Melbourne, 1944: 3; Tree, *op. cit.*: 86–88.
4. John Gould to Sir William Jardine, 30 April 1838, quoted in Tree, *ibid.*: 91–92.
5. John Gould to the Chair of the Scientific Committee of the Zoological Society of London, 20 June 1838, quoted in *ibid.*: 93.
6. Quoted in *ibid.*: 97–98.
7. Elizabeth Gould to Mrs Coxen, 8 October 1838, in Chisholm, *op.cit.*: 33. Original letters can be found on Mitchell Library microfilm reel no. CY1626, available at the National Library of Australia. Originals held by the State Library of New South Wales at ML A174.
8. *ibid.*: 34.
9. Extract from Charles Darwin's *Beagle* diary, reproduced in *Charles Darwin: An Australian Selection*, National Museum of Australia, Canberra, 2008: 68.
10. Lady Franklin to Sir John Franklin, in George Mackaness, *Some Private Correspondence of Sir John and Lady Jane Franklin, (Tasmania, 1837–1845)* vol. I, D.S. Ford, Printers, Sydney, 1947: 38.
11. Sir John Franklin led a final expedition aboard the *Erebus* to the Arctic in 1845. He was 59 years old and died in June 1847, within sight of the North-West Passage, when the ship stuck fast in the ice. None of the expeditioners survived but Lady Franklin pressured the British Government to mount a search for her lost husband and his men, which uncovered relics demonstrating that Franklin had indeed discovered the North-West Passage.
12. Lady Franklin's intrepid nature and intellectual curiosity is explored in Penny Russell, *This Errant Lady: Jane Franklin's Overland Journey to Port Phillip and Sydney, 1839*, National Library of Australia, Canberra, 2002.
13. The Port Davey region is now part of the Tasmanian Wilderness World Heritage Area.
14. Elizabeth Gould to Mrs Coxen, 10 December 1838, in Chisholm, *op. cit.*: 38.
15. Elizabeth Gould to Mrs Coxen, 10 December 1838, in *ibid.*: 37.
16. Elizabeth Gould to Mrs Coxen, 3 January 1839, in *ibid.*: 41.
17. Elizabeth Gould to Mrs Mitchell, 4 January 1839, in *ibid.*: 45–46.

CHAPTER 6
IN THE FIELD: TASMANIA, NEW SOUTH WALES & SOUTH AUSTRALIA

After arriving in Hobart, John Gould lost no time before setting forth to capture as many specimens of Australian birds as he could. At first he could do this without going far afield, obtaining a number of species in the streets of Hobart itself. Very soon, though, he began to range beyond the confines of the town into the country around the Derwent River and Macquarie Plains, often taking his son John Henry with him, along with John Gilbert and James Benstead. Collecting there yielded a rich harvest: most of the species Gould collected in Tasmania were taken within 80 kilometres of Hobart, with many obtained on the steep slopes of Mount Wellington, which loomed over the town.[1]

Gould eagerly took up the opportunity presented by Lady Franklin of a voyage around the southern and western coasts of Tasmania to Port Davey, leaving Hobart on 10 December 1838. Anticipation turned to frustration as bad weather hindered the government schooner's progress, although Gould was able to land and collect at the Actaeon Islands, and Green and Bruny islands. The party got no further than Recherche Bay, where they were, in Gould's words, 'imprizoned'. Impatient to work in further pastures, although he had collected a number of seabird species, including a snowy white petrel that he had particularly coveted, Gould asked if he could return to Hobart, as he had only allocated a month of their stay for Tasmania and there was the rest of Australia yet to be investigated.[2] Lady Franklin, however, was not prepared to give in so soon and insisted that they stay a week longer in the hope that the wind turned in their favour. Gould grumbled to his wife in a letter: 'This is a cold and cheerless place', a charnel-house for whales, with the beach strewn with their skeletons, 'many of which are putrid'. Lady Franklin's determination, nevertheless, had to give way to her husband's wishes: after a fortnight, the party was 'summoned back to Hobarton by Sir John'.[3]

Gould remained in Hobart over the Christmas and New Year period. Elizabeth remarked in a letter to her mother on 3 January 1839 that, while the climate there in summer was 'not very hot usually', on Christmas Day she had 'opened the window a little' and 'In came such a hot puff, as from the mouth of an oven, that down went the window in a moment'.[4] Barely had 1839 begun than Gould and Gilbert set off again overland for Launceston and the islands in Bass Strait.

They reached Launceston by 5 January, collecting specimens on the way, including 'a splendid Eagle' killed by Gould at Perth.[5] On Isabella Island, he killed two Cape Barren geese, and gathered black swan eggs on Flinders Island.[6] Gould had already begun to discover the inestimable value of Aboriginal people in uncovering the natural wonders of Australia and to respect their profound connection to country. From then on, Gould always recorded the Aboriginal name for a bird or mammal, as well as the one derived from Linnaeus' system of binomial nomenclature. The trip to Flinders Island would have been a great success but for one appalling incident that completely rattled the normally robust Gould. He related the story in a letter to Elizabeth:

> a fatal accident happened to one of the men who shot himself dead by uncautiously pulling the gun from the boat with the muzzle toward his chest, the cock of the gun caught the seat of the boat and all was over with the poor fellow in half a minute. I cannot tell you my Dear Eliza how great a shock I sustained. I have scarcely been myself since and I almost hate the sight of a gun.[7]

Henry Constantine Richter (1821–1902)
Osphranter rufus (Great Red Kangaroo, close-up)
in *The Mammals of Australia*, vol. II, by John Gould
(London: John Gould, 1863)

The great project of collecting and recording the birds and, shortly, the mammals, of Australia could not be achieved, Gould realised, unless he and Gilbert separated and collected individually. He therefore instructed Gilbert to remain in Launceston and then take ship for Western Australia, while Gould returned to Hobart to prepare to tackle the eastern portion of the Australian continent, beginning in Sydney. Gilbert left Launceston on the *Comet* on 4 February, not before receiving a letter from Gould setting out his expectations:

> Your attention will be principally directed to ornithology. You will form as complete a collection of birds as you can during your short stay and those in their various stages of plumage together with their nests and eggs, and skeletons of each form. A label being attached to each specimen denoting the colour of the eye feet and bill (with) the date, and locality in which it was killed and obtain and note down any information respecting their habits and economy …
>
> Secondly your attention will next be directed to Mammalia, one or two of every species of which, even the most common (particularly the Kangaroos) will be desirable … from your having been with me for so long here and knowing precisely my present object and wants you will thoroughly understand how to employ yourself. And I trust you will do so with that assiduity which I am happy to acknowledge has shown itself in your persevering and steady conduct since you have been in this colony …
>
> Your stay at Swan River should not be more than 4 or 5 months but if no vessel is leaving for these colonies you will have no alternative but to join me as early as you can.

Unknown artist
Portrait of Dr George Bennett F.Z.S. 1849
lithograph on paper; 32.2 x 28.0 cm
Pictures Collection, nla.pic-an9455386

> You will of course husband your resources and spend as little money as possible …[8]

Gilbert arrived in Western Australia on 6 March for what would be a stay of 11 months. He was paid a salary of £100 per annum and expenses.

Gould's decision to collect Australian mammals may have been made early in his stay in Tasmania. By his own account, Gould owed a great deal to his experiences with Aboriginal people, who had pointed out to him signs of the presence of mammals and brought them as food to the nightly campfire. Gould also observed mammals himself while resting by a river bank, where platypus swam and echidnas 'came trotting up towards me'. He concluded: 'With such scenes as these

ORNITHORHYNCHUS ANATINUS.

ECHIDNA HYSTRIX.

Unknown photographer
Charles Coxen (1808–1876), Brother of Mrs John Gould
detail from *Portrait and Grave of Charles Coxen*
mounted reproductions of original photographic prints; 14.0 x 18.3 cm
Pictures Collection, nla.pic-vn3796973-s1

left, top
Henry Constantine Richter (1821–1902)
Ornithorhynchus anatinus (Ornithorhynchus)
in *The Mammals of Australia*, vol. I, by John Gould (London: John Gould, 1863)

left, bottom
Henry Constantine Richter (1821–1902)
Echidna hystrix (Spiny Echidna)
in *The Mammals of Australia*, vol. I, by John Gould (London: John Gould, 1863)

continually around me, is it surprising that I should have entertained the idea of collecting examples of the indigenous Mammals of a country whose ornithological productions I had gone out expressly to investigate?'⁹

By mid-February, Gould was off again, this time on the *Potentate* bound for Sydney, to begin his collecting in New South Wales. He only stayed there for three days, in the home of Dr George Bennett, Curator of the Australian Museum. Gould and Bennett became friends and Bennett acted as a collector and as a publication agent for Gould after he returned to England. The Gould family stayed with Bennett whenever they were in Sydney and the Australian Museum was a valuable resource for Gould, providing specimens from which he could make identifications when he had been unable to collect a bird or mammal in the field. Many of these specimens had been collected by Gould's brother-in-law, Charles Coxen.¹⁰

Then Gould set off for one of the major sites for his collecting in New South Wales—the upper Hunter River area, where Elizabeth's brother, Stephen Coxen, lived on his property, Yarrundi (now known as Yarrandi), on the Dart Brook, seven kilometres west of Scone. After a week's delay in meeting up with Coxen in Maitland, Gould made haste after reaching Yarrundi to get out into the countryside. He wrote to Elizabeth on 20 March to report that he was 'in excellent health and the colony agrees with me well'. He had already been to the Liverpool Range and had killed nine lyrebirds. Gould also reported on the drought that was devastating the region: while Stephen Coxen was doing better than his neighbours, he was 'suffering much but the distress is general'.¹¹

By this time, Elizabeth was less than two months from her next baby's birth. Gould was frank about the effect that this interruption would have on his collecting goals, telling his wife that: 'I intend leaving this the first moment I can for Sydney hence to Hobart Town to see you, although it will verily interfere with my pursuits'. He warned her that, owing to the difficulties of making his way across country and then by sea to Hobart, he might not arrive in time for the birth. If not: 'You have my sincere hope and prayers for a safe delivery out of your troubles, and which I fear not will be granted to you, and if I am not with you at the time you will have something to present to me when I do'.[12] Gould was, however, back in Hobart by 23 April. Sir John Franklin reported in a letter to his wife, who was on her own expedition from Port Phillip to Sydney, that: 'Mr Gould has made large additions in N.S. Wales to his collection and therefore in high glee'.[13]

Elizabeth and John Gould's youngest son, Franklin Tasman—who had before his birth been dubbed by Elizabeth's cousin, Mrs Mitchell, a 'little convict'— was born on 6 May 1839 at Government House, Hobart. He was named after the Goulds' host, Sir John Franklin, and Abel Tasman, the Dutch seafarer who had first encountered the island that now bears his name. 'Never before have I had so good a time or taken so little medicine', Elizabeth wrote to her cousin several weeks after the birth.[14] The little boy was a great favourite with their Tasmanian hosts; Lady Franklin even asked if he could be given to her to raise but Elizabeth, despite her affection for her hostess, said no.

Within a week of Franklin's birth, Gould left Hobart again, this time for Adelaide. While in Sydney on his way to Yarrundi in March, he had called on Captain Charles Sturt at his property, Varroville, and had discussed with the explorer the possibility of an expedition towards the Murray River. Sturt was to take up the position of Surveyor-General of South Australia in April 1839. Gould travelled on a sheep transport from Launceston to embark on his expedition to South Australia.

Elizabeth, left behind again in Government House, looked after her young baby and worried about her children left behind in England. She wrote to Mrs Mitchell, asking about Louisa's teething and the inevitable question from a distant parent as to whether the children remembered them: 'How are the other dear children? Do they appear to retain any recollections of Mama and Papa?'[15] Elizabeth also declined to take an active part in a ball held at Government House to mark the Queen's birthday and to celebrate Lady Franklin's return from her expedition, asking her cousin: 'Can you imagine such a shy, reserved being as myself frisking about in the midst of 200 people? Don't fancy I attempt dancing— no, no. I am well content to look on'.[16]

Having landed in South Australia at Holdfast Bay, where the sheep and horses were thrown from the transport into the water and had to swim for shore, Gould went on to Adelaide. Three years after its foundation, Adelaide was 'a chaotic jumble of sheds and mud huts, with trees growing here and there in the newly marked-out streets and squares' of Colonel William Light's plan for the city. As in Hobart, Gould found the city streets alive with birds, and the rivers and marshlands around the city promised even more. But protocol demanded that he present his letters of introduction to Governor George Gawler before setting about collecting again. Gawler was able to suggest people who could assist Gould, including one who became a collector for him, Frederick Strange, who worked for the Survey Department. Gould was able to go with Strange on official surveys, including to Angle Vale on the Gawler River. They also explored the rich habitats provided for wildlife by the mangrove swamps of Port Adelaide.[17]

On 17 June, Gould set out with Captain Sturt for the Murray Scrubs (now known as the Mallee). Gould was able to strike out on his own with a cart and a small party towards the Murray River but lack of water among the dense eucalyptus scrub that bordered the river drove him back to the ranges. He stayed a week there, making 'daily visits to this rich arboretum'. He could have stayed indefinitely, 'but, alas! our provisions failing, we were obliged to retrace our steps'.[18] Gould arrived back in Adelaide on 10 July, before setting off for a brief visit to Kangaroo Island, returning to Adelaide and shipping for Hobart on the *Katherine Stewart Forbes* on 19 July.

Unknown artist
Captain Charles Sturt 1895
wood engraving on paper; 15.2 x 10.3 cm
Pictures Collection, nla.pic-an9941030

The time had come to pack up the family and possessions and move on, to the mainland. The Goulds left behind them in Hobart several good friends, including Sir John and Lady Franklin, and the Reverend T.J. Ewing and his wife, Louisa. Ewing became a lifelong correspondent and an active promoter of Gould's publications. Gould not only left behind friends in Hobart but also capital, having made property investments while he was there.[19]

Elizabeth's last letter from Hobart paid tribute to her hosts at Government House: 'I cannot here refrain from alluding to the very great kindness we have experienced from Lady and Sir John Franklin and which will always remain deeply impressed on our memory'. She was on her way to be reunited with her brothers at Yarrundi, where the Goulds would spend several months. By now, Elizabeth had learned that her husband would not stay by her side but that the results of his collecting efforts would compensate for his absence:

> John, of course, will not remain there long, but be wandering off in some direction, though at present his plans on that point are not fixed. He has accomplished an immense deal during the past year and will be able on our return to produce much interesting information.[20]

On 20 August, John and Elizabeth Gould, their sons, John Henry and Franklin, and James Benstead and Mary Watson boarded the *Mary Ann* bound for Sydney and the next stage of their Australian journey.

ENDNOTES

1. Gordon C. Sauer, *John Gould, the Bird Man: A Chronology and Bibliography*, Henry Sotheran Limited, London, 1982: 104; Ann Datta, *John Gould in Australia: Letters and Drawings*, Miegunyah Press, Carlton, Victoria, 1997: 113.
2. George Mackaness, *Some Private Correspondence of Sir John and Lady Franklin*, Sydney, 1947: 49.
3. Sauer, *op. cit.*: 106.
4. Elizabeth Gould to Mrs Coxen, in Alec H. Chisholm, *The Story of Elizabeth Gould*, Hawthorne Press, Melbourne, 1944: 42.
5. John Gould to Elizabeth Gould, 8 January 1839, in Sauer, *op. cit.*: 107.
6. *ibid.*:108.
7. John Gould to Elizabeth Gould, 20 January 1839, in *ibid.*: 109.
8. John Gould to John Gilbert, 29 January 1839, in Datta, *op. cit.*: 115–116.
9. *ibid.*: 120–121.
10. Glenn and Jillian Albrecht, 'The Goulds in the Hunter Region of N.S.W., 1839–1840', *Naturae*, no. 2, August 1992, Centre for Bibliographical and Textual Studies, Monash University, Clayton, Victoria: 2.
11. John Gould to Elizabeth Gould, 20 March 1839, in Sauer, *op. cit.*: 111.
12. John Gould to Elizabeth Gould, 20 March 1839, in *ibid.*: 111.
13. Sir John Franklin to Lady Franklin, 26 April 1839, in Mackaness, *op.cit.*: 68.
14. Elizabeth Gould to Mrs Mitchell, 30 May 1839, in Chisholm, *op. cit.*: 64.
15. Elizabeth Gould to Mrs Mitchell, 28 May 1839, in *ibid.*: 63.
16. Elizabeth Gould to Mrs Mitchell, 30 May 1839, in *ibid.*: 65.
17. Datta, *op. cit.*: 124–125.
18. *ibid.*:126; Letter from John Gould to Editor, *Annals of Natural History*, 28 September 1839, in Sauer, *op. cit.*: 115.
19. Datta, *op. cit.*: 127.
20. Elizabeth Gould to Mrs Coxen, 20 August 1839, in Chisholm, *op. cit.*: 67.

CHAPTER 7
AMONG THE MENURAS IN NEW SOUTH WALES

After a 'rather rough passage' from Hobart, the Gould party arrived in Sydney on 3 September 1839. Elizabeth admired the prospect of the town of Sydney, with its 'picturesque rocky points studded with villas in the suburbs and well built houses in the centre'. When the ship docked, the energetic John Gould disembarked straight away and brought back on board with him Elizabeth's brother, Stephen Coxen, who met her 'with much kindness' and questioned her about all their friends and relations back in England.[1]

The family went to stay with Dr George Bennett and his wife. Elizabeth resumed some of the social life she had first experienced in Hobart, as the Goulds were invited to a ball at Government House, in Sydney. This time, though, she was not permitted to sit to one side, as she had done in Hobart, or claim the need to economise as a reason for not attending. Her brother, clearly determined that his sister should enjoy herself and look her best, 'took me to a shop and did all that was needful *and more*, and called in the evening just before we went purposely to see if I made up well, as he called it'. Coxen *did* approve and told his sister that: 'if John went off the hooks I should soon pick up a capital match'. Governor Sir George Gipps and Lady Gipps 'were very polite and kind', reported Elizabeth, although 'tired enough I

Charles Hullmandel (1789–1850), after Augustus Earle (1793–1838)
Government House and Part of the Town of Sydney c.1830
from *Views in New South Wales and Van Diemen's Land: Australian Scrap Book*, part 2 (London: J. Cross, 1830)
lithograph on paper; 28.0 x 37.5 cm
Pictures Collection, nla.pic-an6065444

was'—unsurprisingly, as she had given birth only four months before.[2]

Elizabeth also strolled about the town with Mrs Bennett, wandered in the Botanic Gardens and noted the quality of food and the prices in

Henry Curzon Allport (1788–1854)
George Street, Sydney, Looking South 1842
watercolour on paper; 33.9 x 50.2 cm
Courtesy State Library of New South Wales

the shops. Fruit was good and plentiful but the prices of staples, such as flour, potatoes, butter, meat, sugar and tea, fluctuated in line with their 'scarcity or abundance', while clothing could be 'absurdly dear'.[3] She enjoyed her brother's company and looked forward to being able to sit down together for a private chat, as 'when he calls at Mr. Bennett's I cannot very well talk of all our family affairs so fully as we shall do' once they arrived at his property, Yarrundi. Seeing one family member though, while satisfying, could not assuage her yearning for those in distant England: 'Oh, my dear Mother, how happy shall I be if permitted to see you once more and my dear children'.[4]

On 14 September 1839, the Gould party and Stephen Coxen left Sydney for Newcastle by steamer, arriving there on 15 September. Coxen went straight on to Yarrundi to get the place ready for their stay, while the Goulds remained behind for a week to explore Newcastle and the islands in the Hunter River. Elizabeth even camped out for a night under canvas on Mosquito Island, 'in the midst of the bush where nature appeared in her wild luxuriance' now that the long drought had broken. She sketched birds and plants and wrote a diary, and John collected specimens, including the regent bowerbird and the figbird—the breeding season had just begun and the birds were in fine feather.[5] John also had his first sighting of a wattled talegalla (brush turkey) on Mosquito Island.[6]

After Newcastle, the Gould party's next destination en route to Yarrundi was Maitland. At 8 am on 22 September 1839, they embarked on the first

top
Botanical studies from *Mrs Elizabeth Gould's Collection of Drawings of Australian Plants, Flowers and Foliage,* **1838–1842**

Courtesy State Library of New South Wales

above
Diary of Elizabeth Gould, 20 August to 29 September 1839

Courtesy State Library of New South Wales

locally built ocean-going paddle steamer in the Australian colonies, *William the Fourth*. The paddle steamer could not travel beyond Morpeth, so the party endured a bone-shaking, five kilometre or so trip in a light cart to Maitland, where they arrived at 2 pm and took lodgings at Cohan's Inn.[7]

From 23 to 28 September, John and Elizabeth spent every day collecting and drawing around Maitland, venturing into the dense bush near the Hunter River, resting in the evening and strolling on the racecourse in the early morning to the sight and sounds of rainbow lorikeets and noisy friarbirds. On 29 September, they left Maitland for a two-day trip by horse and cart, with their luggage hauled by bullock dray, staying overnight at Patrick's Plains, near Singleton. Their tiring journey came to an end when they arrived at Yarrundi on 30 September, where they were to stay for four months. Elizabeth could now settle down with her family, including her brother, as she drew and painted. She was, however, realistic about the fact that John would not spend much time with them: he was anxious to be out and about in the rich habitats of the Upper Hunter district, along with his Aboriginal companions, Natty and Jemmy, who had helped him on his previous visit.[8] One of John's major goals was to obtain some lyrebird eggs and observe their nests. In a letter to Sir William Jardine, he wrote that he hoped to 'spend a week among the *Menuras* ... a cheerful active bird singing and mocking all the birds of the forest'.[9] Gould spent two months there, collecting new species and, aided by Natty's skill in climbing trees, numbers of eggs. He did not, however, manage to observe the nesting habits of lyrebirds.[10]

Several years later, while drafting instructions for John Gilbert for his second trip to Australia, Gould recalled occasions where Natty and Jemmy and other Aboriginal people had helped him; and he encouraged Gilbert to seek their help again:

The great walleroo could be procured by employing the Yarrundi natives (Coxen's). I found them on the hills in front of Mr. Coxen's house, and Natty pointed out a hill close to the cedar brush at the Liverpool Range where they are abundant. Should you visit the Upper Hunter district it would perhaps be well to spend a week in getting as many specimens of the walleroo as possible, always securing the services of Natty and Jemmy, my faithful companions. Mr. Coxen would doubtless lend you a horse and cart and send you and your traps to the range where you would take up your quarters under the very hills on which the animal is found. The rock kangaroo, of which I should like several specimens, is also abundant here. The brushes are not worth hunting for the birds.

You will also in this district procure the rat kangaroo (*Bettongia lesueuri*) with the long white tip to its tail—not the common one found about Yarrundi—but if you call Natty's attention to the one we so often saw on all the low grassy hills immediately adjoining the range he will recollect. I, unfortunately, did not obtain it.

You will, of course, collect every species of mammal in all the districts you visit. Pray try to get a little mouse (somewhat larger than the common European one) with a short blunt head, and which Jemmy caught for me from a hole in the ground in the bush close to Mr. Coxen's garden gate at Yarrundi. There are six or eight opossums in this district, four of which are only found in the brushes. Be sure to get the large great brush opossum with short ear (*Trichosurus caninus*) found in the hollows of the trees.[11]

In mid-November 1839, Gould and his party of six men, including Natty and Jemmy, began a

Henry Constantine Richter, lithographer (1821–1902) and Gabriel Bayfield, colourist (1781–1871), after Elizabeth Gould (1804–1841)
Talegalla lathami (Wattled Talegalla)
in *The Birds of Australia*, vol. V, by John Gould (London: John Gould, 1848)

two-month expedition to the Liverpool Ranges and beyond to the Namoi and Mokai rivers, into grassland rather than bushland, penetrating over 240 kilometres down the Namoi River towards the Brigalow scrub country and ranging over 640 kilometres from the coast. Recent rain had created abundant feed for birds, many of them new to Gould, including the harlequin bronzewing, crimson chat, white-backed swallow, red-backed kingfisher, black-eared cuckoo, Australian egret and striated grass-wren. He also encountered vast flocks of the little green-and-yellow 'warbling grass parakeet' that Aboriginal people called the 'betcherrygah'. Only one specimen of *Melopsittacus undulatus* (warbling grass-parrakeet) had ever been seen in London and now he saw them everywhere. Gould grasped the potential of their charm as cage birds: they had 'a most animated and pleasing song', and were 'constantly billing, cooing, and feeding each other, and assuming every possible variety of graceful position'. He captured 20 to try to bring them home alive but they did not survive the journey back to Yarrundi. Gould's other brother-in-law, Charles Coxen, had managed to raise some in captivity at his Peel River homestead and gave four of them to Gould, who managed to take two alive to England.[12]

Gould's intensive collecting experience in the interior of New South Wales was a highlight of his Australian experience and yielded a rich bounty of bird and mammal specimens, along with the traditional knowledge about them that Natty and Jemmy shared with him and which he incorporated in the commentary to *The Birds of Australia* and *The Mammals of Australia*. Elizabeth, meanwhile, occupied herself at Yarrundi by sketching bird and animal specimens and plants. On 5 December 1839, Yarrundi was visited by a small party from the United States Exploring Expedition, led by Charles Wilkes. One of the visitors, a Dr Charles Pickering, wrote in his journal about Yarrundi and described Elizabeth at her work:

> Mr. Coxen received us very politely and introduced us to his sister Mrs G., to whose talent and industry the world is indebted for the celebrated Ornithological Illustrations. I had the pleasure of seeing the lady at her pencil, and was surprised at the rapidity of her execution.[13]

The Gould party returned to Sydney in February 1840. Gould then set off to visit Bong Bong and Moss Vale, Berrima and Camden, south of Sydney, and travelled the bush roads in the Illawarra district as far as Wollongong, collecting in the area until March. There he enlisted the aid of Charles Throsby of Throsby Park, now a historic site on the outskirts of Moss Vale. Throsby was the nephew and heir of the property's founder, also named Charles Throsby, who had explored the area and was also one of the first settlers in the Illawarra. He had been granted land at Throsby Park by Governor Macquarie in 1820 and 1821, but financial problems led him to commit suicide in 1828.[14]

Throsby had assisted Gould during his stay in the area by making available one or more of his Aboriginal employees to help him with his collecting. Gould, in his turn, confidently recommended that his collector, John Gilbert, should also take advantage of Throsby's help when working in the area. Gould's instructions to Gilbert reflect his own experiences in the Illawarra and the help he had been given there by Throsby and his Aboriginal servants:

> At Illawarra I also saw, but could not procure, a small mouse-like animal among the leaves on the hills. The *Halm. ruficollis* (Warroon of the natives) is also very abundant on the rise of the hills by the side of the bush road between Wollongong and Bongbong. This kangaroo, of which I want good specimens, and particularly a skeleton of an adult male and crania of both sexes, may be procured by paying a visit to Mr. Throsby, at Bongbong, who will send his Tommy or some other native out with you. Bongbong is near Berima and can be reached either by going in the mail cart from Sydney, or by walking over with native guides from Wollongong. The koala or monkey is also common on this road, and in Throsby Park the grey magpie *Strepera*, which I want.[15]

Gould mentioned in his instructions that: 'Both the menura and the talegalla are abundant in some parts of the Illawarra district, but they are scarce near Wollongong'. His failure to find a lyrebird's nest still nagged at Gould and he hoped that Gilbert would succeed where he had not. Gilbert was told by Gould to rely on local Aboriginal knowledge in his quest for the nest: 'you must not fail to gain every information from the natives as to the structure and situation of the nest, number of eggs &c., &c. Offer high rewards for the eggs, but you need not spend time in trying to shoot the birds'. He also wanted Gilbert to clear up a mystery: 'When at Bongbong, the natives also told me of a large blue-grey kangaroo which they sometimes killed. I could never make out what it is unless it be *robustus*. Make enquiry'.[16]

Gilbert was not the only person commissioned by Gould to collect for him in the Australian colonies. Others included Dr George Bennett, Johnson Drummond, Frederick Strange, Charles and Stephen Coxen, and John McGillivray. In addition, soldiers and colonial administrators, and sailors on naval and

Henry Constantine Richter, lithographer (1821–1902), and Gabriel Bayfield, colourist (1781–1871), after Elizabeth Gould (1804–1841)
Menura superba (Lyre Bird)
in *The Birds of Australia*, vol. III, by John Gould (London: John Gould, 1848)

merchant vessels learned to send or bring specimens of unusual Australian birds or mammals to Gould in London. Members of the scientific community in Australia also kept a lookout for possible new species. On 11 April 1848, a teenage girl from Port Macquarie, Annabella Innes, mentioned in her diary that: 'Mr Wilson shewed us a bird he had got at the Macleay and thinks is the first of its kind that has been shot, as they are very rare and shy. He intends to send this specimen home to Mr Gould, the famous ornithologist'.[17]

Annabella Innes (later Boswell) was the niece of Margaret Innes, a daughter of Alexander Macleay, of Elizabeth Bay House, who was Colonial Secretary of New South Wales, a Fellow of the Royal Society of London and the leading instigator in the creation of the Australian Museum, in Sydney. John and Elizabeth Gould had visited Macleay at his home and Elizabeth had sketched a tame brush turkey there that, not long afterwards, met with a sad fate: it mistook its own reflection in a well for that of another bird and, attacking it, fell in and drowned. George Bennett duly prepared it and Gould took it back with him to London.[18]

In her last letter from Australia to her mother, in September 1839, Elizabeth had summed up the results of their stay in the colonies: 'We are as successful here as we can possibly expect'. John Gould had invested £2000 in Tasmania but by far the largest investment had been in added knowledge and in the sheer number of specimens collected to prepare the publications they would produce back in England. Elizabeth was confident that these would succeed: 'I think the great mass of information John has obtained cannot fail to render our work interesting to the scientific world'.[19] On 9 April 1840, John and Elizabeth Gould, their sons, John Henry and Franklin, and Mary Watson and James Benstead left Sydney aboard the *Kinnear*, bound for home.

ENDNOTES

1. Elizabeth Gould to Mrs Mitchell, 4 September 1839, in Alec H. Chisholm, *The Story of Elizabeth Gould*, Hawthorne Press, Melbourne, 1944: 68–69.
2. *ibid.*: 70.
3. *ibid.*: 71.
4. *ibid.*: 73.
5. Glenn and Jillian Albrecht, 'The Goulds in the Hunter Region of N.S.W., 1839–1840', *Naturae*, no. 2, August 1992, Centre for Bibliographical and Textual Studies, Monash University, Clayton, Victoria: 7–9.
6. Isabella Tree, *The Bird Man: The Extraordinary Story of John Gould*, first published 1991, this edition Ebury Press, London, 2004: 151.
7. Albrecht, *op. cit.*: 10.
8. *ibid.*: 10–13.
9. Quoted in Tree, *op. cit.*: 153.
10. *ibid.*: 154. Tree cites a number of new species collected by Gould during this phase of the expedition.
11. H.A. Longman, 'John Gould's Notes for John Gilbert', *Memoirs of the Queensland Museum*, vol. VII, Part IV, issued 19 December 1922: 292–293. The original notes are held in Papers of John Gould 1838–1884, National Library of Australia, Manuscripts Collection, MS 587.
12. Tree, *op. cit.*: 155–160; Gordon C. Sauer, *John Gould, the Bird Man: A Chronology and Bibliography*, Henry Sotheran Limited, London, 1982: 117–118.
13. Quoted in Albrecht, *op.cit.*: 18.
14. Vivienne Parsons, 'Throsby, Charles', *Australian Dictionary of Biography*, vol. 2, Melbourne University Press, 1967: 530–531.
15. Longman, *op. cit.*: 292.
16. *ibid.*: 292.
17. *Annabella Boswell's Journal: An Account of Early Port Macquarie*, edited by Morton Herman, first published Angus and Robertson, Sydney, 1965, this edition 1981: 163.
18. Sauer, *op. cit.*: 118.
19. Elizabeth Gould to Mrs Coxen, 13 September 1839, in Chisholm, *op. cit.*: 74.

CHAPTER 8
EXPERIENCING LOSS

John Gilbert had parted from John Gould in Tasmania in February 1839, remaining in Western Australia until January 1840. Collecting in the west presented some problems. Although he had been 'in the Interior as far as any Europeans have yet settled', Gilbert was hampered in his efforts to collect birds, insects, shells and animals by unrest between Aboriginal people and settlers over 'several frightful murders on the white people, who to punish them killed several of the Blacks in return'. The danger this presented, tribal warfare as well, had 'crippled' his work, although he 'still succeeded in obtaining many Birds not found at Perth'.[1] Leaving the west to rejoin Gould in Sydney also proved difficult, as few ships called at this remote place. When the captain of one that did call in was unable to take on any more passengers and was even unsure if he would be proceeding to Sydney, Gilbert had to remain until another ship appeared. He poured out his plight to Gould:

> I have now been here 6 Months instead of as proposed by you 4 or 5 and as forseen by me (on hearing of the uncertainty of vessels dropping in during the winter and spring months) my cash is nearly expended, for I have

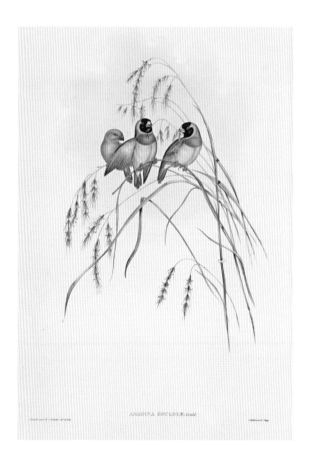

Henry Constantine Richter, lithographer (1821–1902), and Gabriel Bayfield, colourist (1781–1871), after Elizabeth Gould (1804–1841)
Amadina gouldiae (Gouldian Finch)
in *The Birds of Australia*, vol. III, by John Gould (London: John Gould, 1848)

been under the necessity of purchasing cloaths and shoes, (which I would not have done from their being so expensive but I was obliged from sheer necessity), and I need not tell you how destructive the Bush is to cloathing, and much more so to me here than when in V.D. Land, this with Powder and Shot has at least taken away (all but) enough to last me a month or six weeks. I shall be under the necessity of borrowing if I am kept here much longer.[2]

Gilbert was even more frustrated when he finally reached Sydney on 3 May 1840 to find that Gould and his party had sailed for England three weeks before. He complained by letter to Gould that: 'I must say I thought you would not have left before the time expired which I stated you might expect me viz. all April'. He was also disappointed in the haul of natural history specimens that he had accumulated in Western Australia. Gilbert apparently felt that he had to justify to Gould 'so poor a collection of Natural History', but

> I must inform you that Western Australia is wanting in extraordinary novelties … as compared with other parts of the Continent … You must remember I had none of your advantages, what could I not have done with your Bullock Cart Horses and Men? The whole of my collection has been formed by myself unaided by a single person during my whole stay in the Colony.[3]

After a frustrating few weeks in Sydney, Gilbert headed for Port Essington, in northern Australia, leaving on 15 June 1840 on a ship sent with relief supplies to a small settlement that had been devastated by a hurricane. Despite the extreme heat, the landscape with its bird and animal life reduced severely by the hurricane, and ant attacks on his collected specimens, Gilbert found he could move about freely, as the local Aboriginal people were friendly and posed no threat to him. As a consequence, he was able to make journeys of up to ten days into remote northern Australia. He also went by boat to the islands of Van Diemen Gulf, where he discovered 90 new bird species, including a beautiful little finch and a remarkable mound-building bird, the jungle-fowl (mallee fowl). By March 1841, Gilbert was ready to ship for England, after an absence of three-and-a-half years, and returned via Singapore in September.[4]

While Gilbert was collecting in northern Australia, the Goulds arrived back in London in August 1840, after a four-month voyage. Elizabeth was at last reunited with her mother and the beloved children she had left behind, and introduced them to their 'little Tasmanian' brother, Franklin. The ornithological production line at 20 Broad Street resumed and Elizabeth began work on the plates for *The Birds of Australia*, the first part of which was scheduled for publication in December. John Gould was busy with visits to collectors and collections to research his publications, especially a monograph he was writing on the kangaroo. He also showed off some of his ornithological treasures, including 'his pretty singing New South Wales parrots', *Melopsittacus undulatus*, more widely known today as budgerigars and destined to become the most popular cage bird in Europe.[5]

Then disaster struck. Elizabeth died a year after the couple had returned from Australia, at 9 am on 15 August 1841 at the age of 37, at Egham on the Thames, where the family was staying to take advantage of the fresh air. After the birth of an eighth child, Sarah (known as 'Sai'), she contracted puerperal fever, the scourge of childbirth in the days before doctors were aware of the necessity for hygiene. The partnership of Elizabeth and John Gould that had set him on the path to fame and financial success was at an end. Elizabeth was buried in Kensal Green Cemetery, in London. A letter from Lady Franklin to Elizabeth, written

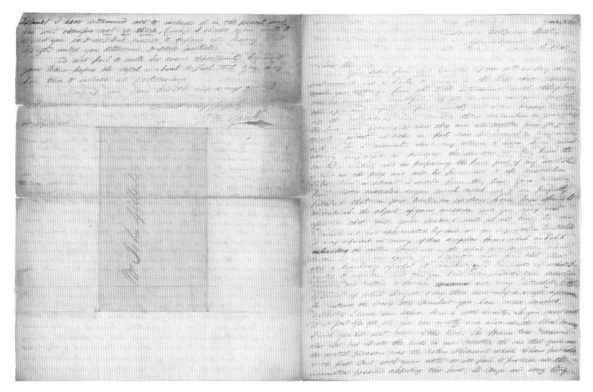

Letter from John Gould to John Gilbert, 16 November 1840

Papers of John Gould 1838–1884, Manuscripts Collection, nla.ms s587-42

in January 1840, testifies to the depth of their friendship: 'I have often thought of you & wished you back again—I fear you are too useful to your husband to be able to flatter myself with this hope'. Now her hopes of seeing Elizabeth again were dashed forever.[6] Another friendship was severed by Elizabeth's death; when Mrs Benstead, who as Mary Watson had accompanied Elizabeth to Australia as her companion, was told of her death by her new employer, Governor George Grey, she 'cried bitterly'. Grey told Gould that Mary Benstead 'still talked frequently and fondly of her former mistress'.[7]

In true Victorian manner, Gould did not show his emotions in public and turned to his work to assuage his grief and anxiety about raising six children as a sole parent. The arrival of John Gilbert from Australia six weeks after Elizabeth's death was a valuable diversion from his private troubles and Gould flung himself into an examination of Gilbert's rich haul of bird and animal specimens.[8] Every now and again, though, he allowed his feelings to show. A poignant request to a subscriber asking for a comment on Elizabeth's contribution to Part 5 of *The Birds of Australia* shows how much he wanted to hear her work praised and how aware he was that it would not continue: 'I shall be glad of a line saying how you like the present part; almost the last of the work of my Dear and never to be forgotten partner'.[9]

Three years later, when Part 15 of *The Birds of Australia* was published, Gould paid his late wife the highest tribute he could offer. He named the lovely little multicoloured finch collected by Gilbert in northern Australia—*Amadina gouldiae*, the Gouldian finch—after Elizabeth, the only time he ever applied either his name or hers to a bird.[10] Opposite the exquisite lithographic plate depicting the bird are his words:

> It is therefore with feelings of no ordinary nature that I have ventured to dedicate this new and lovely bird, to the memory of her, who in addition to being a most affectionate wife, for a number of years laboured so hard and so zealously assisted me with her pencil in my various works, but who, after having made a circuit of the globe with me, and braved many dangers with a courage only equalled by her virtues, and while cheerfully engaged in illustrating the present work, was by the Divine will of her Maker suddenly called from this to a brighter and better world; and I feel assured in dedicating this bird to the memory of Mrs. Gould, I shall have the full sanction of all who were personally acquainted with her, as well as those who only know her by her delicate works as a artist.[11]

After Gould's death 40 years later, one of his friends, author and journalist Blanchard Jerrold, referred to Elizabeth as 'a devoted Priestess of Nature' and asserted that Gould was keenly appreciative of her contribution to his success:

> The loving, skilful hands of Mrs. Gould were at work, painting the birds which she and her husband so passionately studied together, through trials and perils innumerable, for many a long year … Mr. Gould never failed to tell his friends how deep was his debt of gratitude to the artistic aptitude and courageous devotion of his wife and fellow-traveller. She it was who gave form and colour to his 600 varieties of birds. It would grieve him could he know that this debt of his had been overlooked.[12]

Elizabeth Gould, indeed, proved impossible to replace in one important sphere of Gould's life. While he soon found able and talented illustrators, such as Henry Constantine Richter, William Hart, Joseph Wolf and others, to take her place at the lithographic stones, Gould never found another marital partner. He did have the friendship of several natural historians, including Prince Charles Lucian Bonaparte, who was said to be 'always with Gould' when he was in England. In a letter to Bonaparte, a mutual friend, John Edward Gray, relayed some gossip he had heard about Gould's 'wife-hunting'. According to Gray, Gould had 'asked at least fifty' women to marry him, without success. Gray continued: 'He is not likely to get so good and useful a wife as his first in a hurry and has made himself rather ridiculous by his proceeding'.[13]

John Gilbert, meanwhile, was keen to return to Australia and Gould commissioned him again as a collector. Gilbert left England on 2 February 1842 and arrived in Fremantle, Western Australia, on 17 July, with a salary of £100 per year, ten per cent on Gould's book sales and another ten per cent on the sale of specimens that Gould did not need for his own work.[14] Gilbert collected in Western Australia for 17 months, venturing as far afield as the Abrolhos Islands. Many of the 432 bird and 318 mammal specimens, as well as reptiles and plants collected in remote areas, were new to science, and Gilbert made valuable field notes. He travelled to Sydney in January 1844 and then trekked overland to the Darling Downs, in Queensland.[15]

Gilbert's wanderings meant infrequent communication. After a while, Gould began to fret at not hearing from him. In April 1844, the silence having lasted for 13 months, he wrote to Gilbert of 'some apprehensions for your safety which I would be glad to have allayed by the first opportunity'.[16] When letters from Gilbert finally arrived in August 1844, the relieved Gould wrote back, his words full of concern for his collector, which had extended to contacting the Governor of Western Australia for help: 'Rest assured that by me you are never forgotten … on the contrary your long silence caused me very great anxiety for your welfare and safety; as you will find by my letter to Governor Hutt requesting him to institute inquiries after you'.[17]

In September 1844, Gilbert fell in with the German naturalist–explorer, Ludwig Leichhardt, and his party on its way to Port Essington. Gilbert was soon recognised as having superior bush skills to the rest of the party—including its leader, whose shortcomings he noted in his diary that was discovered many years later among Gould family effects—and as the one who understood how to deal with the Aboriginal people they encountered. It is thus a tragic irony that the one person in Leichhardt's party who best understood Aboriginal people should die at their hands.[18] Probably in retaliation for the possible rape of women by two Aboriginal men in Leichhardt's party,[19] a group attacked the camp near the Gulf of Carpentaria on 28 June 1845. Leichhardt's journal of the expedition related the details of Gilbert's death:

> After dinner, Messrs. Roper and Calvert retired to their tent, and Mr. Gilbert, John, and Brown, were platting palm leaves to make a hat, and I stood musing near their fire place, looking at their work, and occasionally joining in their conversation. Mr. Gilbert was congratulating himself upon having succeeded in learning to plat; and, when he had nearly completed a yard, he retired with John to their tent. This was about 7 o'clock; and I stretched myself upon the ground as usual, at a little distance from the fire, and fell into a dose, from which I was suddenly roused by a loud noise, and a call for help from Calvert and Roper. Natives had suddenly attacked us. They had doubtless watched our movements during the afternoon, and marked the position of the different tents; and, as soon as it was dark, sneaked upon us and threw a shower of spears at the tents of Calvert, Roper, and Gilbert, and a few at that of Phillips, and also one or two towards the fire. Charley and Brown called for caps, which I hastened to find, and, as soon as they were provided, they

discharged their guns into the crowd of the natives, who instantly fled, leaving Roper and Calvert pierced with several spears, and severely beaten by their waddies ... Not seeing Mr. Gilbert, I asked for him, when Charley told me that our unfortunate companion was no more! He had come out of his tent with his gun, shot, and powder, and handed them to him, when he instantly dropped down dead. Upon receiving this afflicting intelligence, I hastened to the spot, and found Charley's account too true. He was lying on the ground at a little distance from our fire, and, upon examining him, I soon found, to my sorrow, that every sign of life had disappeared. The body was, however, still warm, and I opened the veins of both arms, as well as the temporal artery, but in vain; the stream of life had stopped, and he was numbered with the dead ... The spear that terminated poor Gilbert's existence, had entered the chest, between the clavicle and the neck; but made so small a wound, that, for some time, I was unable to detect it. From the direction of the wound, he had probably received the spear when stooping to leave his tent ... I interred the body of our ill-fated companion in the afternoon (of 29 June), and read the funeral service of the English church over him. A large fire was afterwards made over the grave, to prevent the natives from detecting and disinterring the body.[20]

The year 1845 saw other tragedies befall Gould's associates: Gould's brother-in-law, Stephen Coxen, his finances in ruins, committed suicide in a Sydney hotel on 4 September; Johnson Drummond, engaged by Gilbert to collect for Gould in Western Australia, was killed by Aboriginal people. Collecting in remote Australia carried significant risks: nine years later, another of Gould's Australian collectors, Frederick

Charles Rodius (1802–1860)
Dr Leichhardt, 1846 1847?
lithograph on paper; 14.0 x 8.5 cm
Pictures Collection, nla.pic-an5600270

Strange, was killed by Aboriginal people, on 14 October 1854 on South Percy Island, off Cape Capricorn, in Queensland.[21]

Gould heard of Drummond's death while he was awaiting news from Gilbert, in reality dead for nearly six months. He would not learn the details of Gilbert's fate until a letter from John Roper, another member of Leichhardt's expedition who had been wounded in the same attack, was sent to Gould on 12 May 1846 and reached him in August. Roper

Blacks about to attack Leichhardt's Camp, near the Gulf of Carpentaria, 1845.

wrote of Gilbert: 'As a companion none was more cheerful or more agreeable; as a man, none more indefatigable or more perservering'.[22]

Gould paid the same tribute to Gilbert as he had to Elizabeth: he named an Australian mammal, a potoroo (*Potorous gilberti* or Gilbert's potoroo) in his honour. Gilbert had collected the little animal at King George's Sound in 1840 during his first visit to Australia. The letterpress of *The Mammals of Australia* carries Gould's expression of appreciation of Gilbert's massive contribution to his Australian collections:

> In dedicating it (the potoroo) to the late Mr. Gilbert, who proceeded with me to Australia to assist in the objects of my expedition, I embraced with pleasure the opportunity afforded me of expressing my sense of the great zeal and assiduity he displayed in the objects of his mission; and as science is indebted to him for his knowledge of this and several other interesting discoveries, I trust that … it will not be deemed inappropriate.[23]

After J. Macfarlane (active 1890–1898)
Blacks about to Attack Leichhardt's Camp, near the Gulf of Carpentaria, 1845 **1890s**
from the Australian Explorers series (Melbourne: George Robertson & Co., 1890s)
photo-engraving on paper; 52.5 x 69.0 cm
Pictures Collection, nla.pic-an9025855-4

Henry Constantine Richter, lithographer (1821–1902)
Hypsiprymnus gilberti (Gilbert's Rat-Kangaroo)
from *The Mammals of Australia*, vol. II, by John Gould (London: John Gould, 1863)

Around the same time as Gould learned the details of Gilbert's death, he suffered another loss. In September 1845, several of his most beautiful birds were stolen, including the type specimen of *Emblema picta* (painted finch), a specimen of the paradise parrot collected by Gilbert in 1844,[24] a northern rosella, a budgerigar, two kingfishers and two warblers. Gould's offer of a £20 reward was to no avail and the birds were never recovered.[25]

Despite the tragic events of the early to mid-1840s, success was also a persistent note in Gould's story. He was elected a Fellow of the Royal Society of London in 1843, oversaw the publication of the parts of *The Birds of Australia* and, from 1845, *The Mammals of Australia*, and sold some of his bird specimens from South Australia to the British Museum.[26] His mother-in-law, Mrs Coxen, and Elizabeth's cousin, Mrs Mitchell, took on the caring role for Gould's children, which they had assumed when Elizabeth and John went to Australia. Gould found companions at his club and, for recreation, he went fishing. He was also pondering new works on other bird species—one in particular, hummingbirds, had captured his attention and would be the subject of his next venture.

ENDNOTES

1. John Gilbert to John Gould, Perth, 3 September 1839, in Ann Mozley Moyal (ed.), *Scientists in Nineteenth-century Australia: A Documentary History*, Cassell Australia, Stanmore, New South Wales, 1976: 64–65.

2. *ibid.*: 65.

3. John Gilbert to John Gould, Sydney, 4 May 1840, in *ibid.*: 65–66.

4. Alec H. Chisholm, *Strange New World: The Adventures of John Gilbert and Ludwig Leichhardt*, Angus and Robertson, Sydney, 1941: 52–55.

5. Gordon C. Sauer, *John Gould, the Bird Man: A Chronology and Bibliography*, Henry Sotheran Limited, London, 1982: 120; Isabella Tree, *The Bird Man: The Extraordinary Story of John Gould*, first published 1991, this edition Ebury Press, London, 2004: 157–159.

6. Lady Franklin to Elizabeth Gould, 15 January 1840, Manuscripts Collection, National Library of Australia, MS 587, Folder 2.

7. Alec H. Chisholm, *The Story of Elizabeth Gould*, Hawthorne Press, Melbourne, 1944: 25.

8. Sauer, *op. cit.*: 120.

9. John Gould to Hugh Strickland, 3 November 1841, in Sauer, *op. cit.*: 42.

10. *Strange New World, op. cit.*: 54. Chisholm gives the Gouldian finch a different first name, *Poephila gouldiae*. Other names include *Chloebia gouldiae*.

11. John Gould, *The Birds of Australia*, vol. 15, 1 June 1844, 20 Broad Street, London, 1844.

12. Blanchard Jerrold to *The Times*, published 11 February 1888, quoted in Richard Bowdler Sharpe, *An Analytical Index to the Works of the Late John Gould, F.R.S., with a Biographical Memoir and Portrait*, Henry Sotheran, London, 1893: xii–xiii.

13. John Edward Gray to Prince Charles Lucian Bonaparte, 21 February 1844, quoted in Sauer, *op. cit.*: 577–578.

14. Sauer, *op. cit.*: 122; Penny Olsen, *Glimpses of Paradise: The Quest for the Beautiful Parakeet*, National Library of Australia, Canberra, 2007: 13.

15. Alec H. Chisholm, 'Gilbert, John (1810?–1845)', *Australian Dictionary of Biography*, vol. 1, Melbourne University Press, Melbourne, 1966: 441–442.

16. John Gould to John Gilbert, 18 April 1844, in Olsen, *op. cit.*: 17.

17. John Gould to John Gilbert, 24 August 1844, Manuscripts Collection, National Library of Australia, MS 587/9, Folder 1.

18. *Strange New World, op.cit.*: 268.

19. Tree, *op. cit.*, 187–193.

20. Ludwig Leichhardt, *Journal of an Overland Expedition from Moreton Bay to Port Essington*, T. & W. Boone, London, 1847: 308–311.

21. Sauer, *op. cit.*: 136.

22. Tree, *op. cit.*: 188–189, 191.

23. John Gould, *The Mammals of Australia*, quoted in Kevin Kenneally, 'John Gould: Nature's Illustrious Illuminator', *Landscope*, vol. 20, no. 1, Spring 2004, Western Australian Department of Conservation and Land Management, Perth: 36.

24. Olsen, *op. cit.*: 14.

25. Alec H. Chisholm, 'John Gould's Stolen Birds', *Victorian Naturalist*, vol. LVIII, January 1942: 131–133.

26. Sauer, *op. cit.*: 120.

CHAPTER 9
VICTORIAN ENTREPRENEUR AMONG THE HUMMINGBIRDS

'It was on the 21st of May, 1857, that my earnest day thoughts and not infrequent night-dreams of thirty years were realized by the sight of a Humming Bird', wrote John Gould in his publication on the *Trochilidae*, the hummingbird family. 'The pleasure which I experience on seeing a Humming-Bird is as great at the present moment as when I first saw one'.[1]

The first sight of a hummingbird in flight is an unforgettable experience. A blurred movement among the leaves at first suggests a large insect—a bumblebee or a dragonfly—until you see a long beak protruding above an iridescent body and realise that what you are seeing is a hummingbird. Every metaphor that has ever been used to describe this avian phenomenon then comes to mind: 'flying jewel', 'feathered gem' are only two of many. Ornithologists suffered from the same urge to bestow lapidary terms on hummingbirds. Eminent nineteenth-century naturalist Professor Alfred Newton, describing hummingbirds for *Encyclopedia Britannica*, wrote that: 'there is hardly a precious stone—ruby, amethyst, sapphire, emerald, or topaz—the name of which may not fitly, and without any exaggeration, be employed in regard to humming-birds'.[2]

By his own account, Gould had been obsessed with hummingbirds for three decades, finally seeing one in the wild in Philadelphia while on a visit to North America with his son Charles in 1857. He had begun the publication of his celebrated five-volume *Monograph of the Trochilidae, or Family of Hummingbirds* in 1849, a process concluded in 1861. By the time he died, Gould had amassed a stupendous collection of 1500 mounted and 3800 unmounted hummingbird specimens that, unlike his other bird specimens, he never sold.[3]

Six years before seeing live hummingbirds, Gould had made their beauty widely known in an exhibition of the exquisite tiny birds at Regent's Park Zoo, staged concurrently with the Great Exhibition of 1851 and seen by over 80 000 visitors.[4] Displaying considerable showmanship, Gould mounted the glittering little birds (they retain their colours in death because of the particular anatomy of their feather barbs[5]) in revolving showcases, lit so that their plumage would glisten. On 10 June 1851, the exhibition attracted royal patronage: Queen Victoria enthused in her journal that she could not 'imagine anything so lovely' as the hummingbirds, with 'their variety, & the extraordinary brilliancy of their colours'.[6] Charles Dickens conjured up a vivid metaphor,

William Hart (1830–1908)

Helianthea osculans

from *A Monograph of the Trochilidae, or Family of Hummingbirds,*
Supplement, by John Gould (London: John Gould, 1861)
Courtesy Australian Museum Research Library

above
One of the hummingbird cases that John Gould displayed
at Regent's Park Zoo in 1851

Courtesy Natural History Museum, London

top
Unknown artist

*Interior of the Humming-bird House, in the Gardens of the Zoological
Society* 1852

from *The London Illustrated News*, London, 12 June 1852

'The Tresses of the Day Star', for the title of his *Household Words* article on Gould's hummingbird exhibition.[7] After seeing the display, John Ruskin lamented that he had wasted his life in studying mineralogy and should have studied the lives and plumage of birds instead: 'If I could only have seen a humming-bird fly, it would have been an epoch in my life'.[8] Gould's hummingbird exhibition sealed his reputation as 'one of the greatest ornithologists alive'.[9]

The years between the Goulds' return from Australia, Elizabeth's death in August 1841 and John Gould's triumphant display of hummingbirds at Regent's Park Zoo ten years later had seen the relentless stream of publications from Broad Street continue unabated. The first part of *The Birds of Australia* was issued in December 1840 and thereafter at three-monthly intervals until December 1848; from 1841 to 1851, another eight titles were added to the list, including Gould's monograph on the kangaroo and *The Mammals of Australia*.[10] Gould's collaborator in this great effort was Henry Constantine Richter, a talented lithographer who replaced Elizabeth in December 1841 and learned to replicate the illustrative style that she and John had developed.[11]

Gould was unsentimental: his business acumen trumped nostalgia when it came to capitalising on the natural history resources that he and his collectors had gathered, often at enormous human cost, from the far corners of the world. Apart from his beloved hummingbirds, once he had described and illustrated a bird or animal species, its stuffed body or preserved skin was offered for sale. As the British Museum had already purchased some of Gould's Australian birds in 1841, he confidently offered the rest of his Australian bird collection to the Museum in 1847, for £1000. He was mortified when the British Museum trustees declined the purchase and, 'in a moment of chagrin', he approached another institution, this time in the United States.[12] Gould knew that Dr Thomas B. Wilson of the Academy of Natural Sciences, in Philadelphia, had already purchased a major ornithological collection and was in the market for more to seal his academy's reputation. Gould offered two options: the birds alone for £800 or with the eggs for £1000. Wilson eagerly seized

Unknown photographer
Portrait of Richard Bowdler Sharpe c. 1900
black and white photograph; 10.8 x 8.8 cm
Pictures Collection, nla.pic-vn3798991

Unknown photographer
John Gould's House on Broad Street, London
Courtesy Natural History Museum, London

the opportunity and the deal was struck. Gould's Australian birds and their eggs were sent across the Atlantic. Gould's friend and patron, Sir William Jardine, pleaded with Wilson to reconsider his purchase, on the grounds that: 'such a collection as that you have purchased as the types of an extensive and important work should not leave this country, but should have been taken by some public museum where they could at all times be accessible for study or reference'.[13] Wilson was unmoved by Jardine's appeal: the collection went to Paris for mounting by the Verreaux brothers, who removed Gould's original labels and affixed an abbreviated version of Gould's data to the stands, and arrived at the academy in June 1849.[14] The British Museum, though, had learned its lesson: after this loss to a transatlantic rival—described by Gould's friend and keeper of the British Museum's bird collection, Richard Bowdler Sharpe, as 'nothing less than a national disaster'—it bought all of Gould's remaining bird and egg collections after he died.[15]

Another artist, William Hart, joined Gould's team at 20 Broad Street in 1851. Gould's eldest daughter, Eliza, recalled the daily routine of life in the five-storey Georgian terrace house. The children saw little of their father throughout the day, as they had their meals separately and 'he generally dined at his club when we were little'. However, they 'always saw him in the morning between nine and ten, when he went down to breakfast, and all made a rush at him for a kiss, nearly pulling his head off, he used to say and again when he came up to change his coat for late dinner'. Gould worked away with Edwin Charles Prince, his secretary, and with Richter and Hart 'in his office at the back of the house, busy with his birds and books'. The 'very pretty' drawing room gradually filled up with Gould's hummingbird collection and was not used or needed, 'as we were children and father having no wife did not have ladies to see him, unless it was to see the birds', Eliza wrote.[16]

The Gould's home overlooked a pump in the street. One of the Gould children's 'great amusements' was to watch people coming and going at the pump that dispensed water that 'was considered very good then & people sent jugs to be filled from all around'.[17] On the morning of Saturday 2 September 1854, Gould returned home from a trip out of London.

Asking for some water, he 'was much surprised to find that it had an offensive smell, although perfectly transparent and fresh from the pump'.[18] Fortunately, he did not drink any of it. He discovered that one of his servants had drunk water from the pump and had become violently ill with cholera the day before he came home. She recovered but thousands of people in London did not. Physician John Snow tracked down the source of the cholera epidemic to the Broad Street pump outside the Gould's house, tainted with raw sewage that seeped from cesspits in the overcrowded Soho area.[19]

Personal tragedy was averted then but struck a year later when Gould's eldest son, John Henry, who had accompanied him and Elizabeth to Australia, died of fever at the age of 24 on 4 October 1855 in Bombay, where he was an assistant surgeon for the East India Company.[20] Again, as he had done when Elizabeth died, John Gould took refuge in natural history, writing to his friend, Prince Charles Lucian Bonaparte, to thank him for 'the kind sympathy you have been pleased to express in the heavy loss I have lately sustained in the death of my eldest son'. He attributed his capacity to 'bear it with fortitude' to the knowledge that 'I have allways done my duty to all my children [and to] the pleasing occupation in which I am

Unkown photographer
Lizzie, Louisa and Sai (Gould's Daughters) 1862?
Private Collection

Unknown photographer
Joseph Wolf
Courtesy Natural History Museum, London

engaged … the Science of Nat. History in which we are both so much interested'.[21] Two years later, he travelled with his son Charles to the United States and achieved his hummingbird dream: first seeing a single live hummingbird in Bartram's Gardens, in Philadelphia, and then observing even greater hummingbird riches in Washington: 'I was now gratified by the sight of from fifty to sixty on a single tree'.[22]

Gould's faithful secretary, Prince, had been complaining for years that the Broad Street house, leaving aside its proximity to the cholera-dealing pump, was dark and unhealthy. Gould, obsessed with his work, seemingly impervious to the discomforts and ills experienced by others in his household and understandably reluctant to interrupt his publication schedules by moving house, did not relocate until December 1859. His new address was 26 Charlotte Street, Bedford Square (now 23 Bloomsbury Street), a more respectable neighbourhood than Soho, with the added advantage that it was close to the British Museum.[23] Gould would live there for the rest of his life, overseeing the stream of publications, for which he had also engaged 'one of the most accomplished of bird-painters', Joseph Wolf, who contributed plates to some of Gould's later works, including *The Birds of Asia* and *The Birds of Great Britain*.[24]

While Gould's three daughters, Eliza (Lizzie), Louisa and Sarah (Sai), stayed with their father, his surviving sons, Charles and Franklin, were heading out into the world. Charles, a geologist, had already left England by the time the family moved to Charlotte Street and, by January 1859, he was on his way to Tasmania to undertake a geological survey. In 1866, Gould's youngest son, Tasmanian-born Franklin, graduated as Doctor of Medicine from the University of Edinburgh. On 7 September 1869, Gould's eldest daughter and her mother's namesake, Eliza, married John Muskett, becoming the only one of the Gould children to marry and have children, when her daughter Helen was born on 8 January 1877.[25]

Before he became a grandfather, though, Gould suffered another tragic blow: on 19 March 1873, Elizabeth's 'little Tasmanian', Franklin, died of fever during a journey from Bombay to Suez with Earl Grosvenor, to whom he was personal medical attendant. Franklin was buried at sea.[26] He had been Gould's favourite son and this third blow was the most telling Gould had experienced. Informing her sister Eliza of the dreadful news, Sarah wrote: 'Father … seems sort of broken with it, & says

Unknown artist
Portrait of Elderly John Gould 1870s
sepia-toned photograph; 8.3 x 5.2 cm
Pictures Collection, nla.pic-an24770488

so little, but began to talk this evening of sending for Charlie in case anything happened to him'. Charles returned briefly but left England again forever, eventually settling in South America.[27]

Gould's friendship with Richard Bowdler Sharpe, whom he had met while fishing on the Thames at Cookham in 1862, was invaluable when Gould's dedicated assistant, Edwin Prince, became severely ill in 1872, causing the elderly Gould much anxiety. Bowdler Sharpe later recalled that Gould, by that time 'himself an invalid, used to drive out himself every day to see his old friend and to carry with him everything that he fancied would do good to the sufferer'.[28] Prince died on 16 August 1874.[29] Gould turned to Bowdler Sharpe to help him prepare *The Birds of New Guinea*, the first part of which was published in 1875. Bowdler Sharpe, who became an eminent ornithologist himself, would see the rest through publication until they were completed in 1888.

John Gould died on 3 February 1881, aged 76, and was buried at Kensal Green Cemetery, in London, where his wife Elizabeth had been laid to rest 40 years before. Nearby lay the body of Gould's secretary, Edwin Charles Prince. Lady Franklin, who had befriended the Goulds on the far side of the world in Tasmania, also lay in Kensal Green. Before he died, Gould had chosen his epitaph: his gravestone reads 'Here lies John Gould, the "Bird Man"'.[30]

ENDNOTES

1. John Gould, *Trochilidae*, vol. 3: 131, quoted in Gordon C. Sauer, *John Gould, the Bird Man: A Chronology and Bibliography*, Henry Sotheran Limited, London, 1982: 139; John Gould, *Introduction to the* Trochilidae: (i), quoted in Judith Pascoe, *The Hummingbird Cabinet: A Rare and Curious History of Romantic Collectors*, Cornell University Press, Ithaca, New York, 2006: 42.

2. Quoted in Wilfrid Blunt, *Ark in the Park: The Zoo in the Nineteenth Century*, Hamish Hamilton, London, 1976: 61.

3. Isabella Tree, *The Bird Man: The Extraordinary Story of John Gould*, first published 1991, this edition Ebury Press, London, 2004: 210.

4. Gordon C. Sauer and Ann Datta (eds), *John Gould, the Bird Man: Correspondence, with a Chronology of his Life and Works*, Mansfield Centre, Connecticut, USA, in association with the Natural History Museum, Maurizio Martino, London, c.1998–2006: 4.

5. Pascoe, *op. cit.*: 33.

6. Queen Victoria. Manuscript. 1851. Queen Victoria's Journal. Collection: The Royal Library, Windsor Castle, through the courtesy of Her Gracious Majesty Queen Elizabeth II, entry 10 June 1851, quoted in Sauer and Datta (eds), *op. cit.*: 323.

7. Charles Dickens, 'The Tresses of the Day Star', *Household Words*, no. 65, 1851.

8. Pascoe, *op. cit.*: 43.

9. Tree, *op. cit.*: 235.

10. A full listing of John Gould's major publications, with dates of issue, can be found at Appendix 1.

11. Sauer and Datta (eds), *op. cit.*: 122; Allan McEvey, 'John Gould's Contribution to British Art: A Note on Its Authenticity', *Art Monograph 2*, Sydney University Press for the Australian Academy of the Humanities, 1973: 8–9.

12. Tree, *op. cit.*: 209.

13. *ibid.*: 207.

14. Sauer and Datta (eds), *op. cit.*: 126–127.

15. Tree, *op. cit.*: 208–209.

16. Quoted in McEvey, *op. cit.*: 6.

17. Tree, *op. cit.*: 239.

18. John Snow, quoted in Sauer and Datta (eds), *op. cit.*: 136.

19. Tree, *op. cit.*: 238–244.

20. Sauer and Datta (eds), *op. cit.*: 137.

21. Quoted in Tree, *op. cit.*: 248.

22. John Gould's *Trochilidae*, vol. III, plate 131, quoted in Sauer and Datta (eds), *op. cit.*: 139.

23. *ibid.*: 144.

24. McEvey, *op. cit.*: 9.

25. Sauer and Datta (eds), *op. cit.*: 143, 146, 147, 149.

26. *ibid.*: 147.

27. Tree, *op. cit.*: 279–280.

28. Richard Bowdler Sharpe, *An Analytical Index to the Works of the Late John Gould, F.R.S., with a Biographical Memoir and Portrait,* Henry Sotheran, London, 1893: xxiv.

29. Maureen Lambourne and Christine E. Jackson, 'Mr. Prince: John Gould's Invaluable Secretary', *Naturae*, no. 4, 1993, Centre for Bibliographical and Textual Studies, Monash University, Clayton, Victoria: 17.

30. Sauer and Datta (eds), *op. cit.*: 151; Tree, *op. cit.*: 294.

EPILOGUE
GOULD'S LEGACY IN AUSTRALIA

John Gould's legacy in Australia is threefold: he is regarded as the father of Australian ornithology, as an early prophet of environmental damage, such as species extinction, and as the man whose name was adopted by one of Australia's most enduring nature conservation organisations, the Gould League.

Commentators on Gould's work have acknowledged his crucial importance to Australian ornithology, as well as his contribution to ornithological illustration generally. Gordon Sauer, who devoted a lifetime to the study of Gould and his works, wrote that: 'No other ornithologist has ever exceeded (or will ever exceed) the number of Gould's bird discoveries and the magnitude and splendour of his folio publications'.[1] Allan McEvey asserted that for most of the nineteenth century: 'Gould plates were a dominating factor in British ornithological illustration'.[2] Some may dispute Gould's artistic standing by comparison with other artist–naturalists of the time, such as John James Audubon, but for Australian ornithology, Gould and his artists stood supreme throughout

Detail of the facade of the Natural History Museum, London, which holds some of John Gould's bird collections and the bulk of his correspondence

Courtesy the author

A Series of Gould's Most Attractive Works

from *An Analytical Index to the Works of the Late John Gould, F.R.S.* by Richard Bowdler Sharpe (London: Henry Sotheran and Co., 1893)

the nineteenth century. In terms of knowledge of Australian bird species, Gould is the undeniable leader: he described almost half of Australia's 745 bird species (between 300 and 328 new species), as well as 45 Australian mammals.[3]

A recent book on Australian ornithology has restated Gould's position among the greats of natural history in this country:

> Ornithology in Australia had a strong start with the 19th century work of John Gould and his extraordinary collectors, especially John Gilbert. Gould's *Birds of Australia*, published in fascicles between 1840 and 1848, quickly became the definitive work on the subject, and remained so for the rest of the 19th century. It is still regarded as an extremely important reference in the history of ornithology.[4]

Gould's publications, as well as informing the world of natural science about Australian birds and mammals, also raised awareness of possible extinction. He warned in 1865:

> It may be possible—and, indeed it is most likely—that flocks of Parakeets no longer fly over the houses and chase each other in the streets of Hobart Town and Adelaide, that no longer does the noble Bustard stalk over the flats of the Upper Hunter, nor the Emus feed and breed on the Liverpool plains, as they did at that time; and if this be so, surely the Australians should at once bestir themselves to render protection to these and many other native birds; otherwise very many of them … will soon become extinct.[5]

One of Gould's most popular images in Australia is the depiction of the thylacine (Tasmanian tiger), one of 21 mammal species and 23 bird species to have become extinct since his time. Gould wrote of

Henry Constantine Richter (1821–1902)
Thylacinus cynocephalus (Thylacine, head)
in *The Mammals of Australia*, vol. I, by John Gould (London: John Gould, 1863)

the thylacine that, once Tasmania became 'densely populated … the numbers of this singular animal will speedily diminish, extermination will have its full sway, and it will then like the wolf in England and Scotland, be recorded as an animal of the past'. He advocated the passing of laws to preserve our 'highly singular, and in many cases noble, indigenous animals'.[6] His prediction of species extinction made nearly 150 years ago was all too accurate, as Australia has 'the doubtful honour of leading the world in mammalian extinctions'—a staggering 30 per cent of all endemic Australian species that existed in 1788 have disappeared—and 'the highest number of threatened species on the planet'.[7] Nowadays, Gould's thylacine image is frequently used to call the attention of generations of Australians to the tragedy of species extinction, while works of ornithology, such as Penny Olsen's *Glimpses of Paradise: The Quest for the Beautiful Parakeet*,[8] also show us what we have lost since Europeans settled Australia in the late eighteenth and early nineteenth centuries.

At a time when homesick colonists were beginning to introduce birds and animals from England to their new homeland, Gould warned that this practice could lead to environmental disaster, calling Australians 'doubly short-sighted … for wishing to introduce into Australia the productions of other climes, whose forms and nature are not adapted to that country'.[9] By the time Gould died in 1881, one of the animal species introduced to Australia, the rabbit, was creating an environmental disaster. Australians now understand that introduced species, such as cats, pigs, dogs, horses, camels, foxes, rabbits, hares and numerous bird species, are responsible for the continuing extinction of indigenous birds and mammals.

Generations of Australian school children recognise the name of John Gould because of the Gould League, which began in Victoria in 1909 as the Gould League of Bird Lovers. A year later, a Gould League was established in New South Wales; Tasmania and Western Australia followed in 1920 and 1939, respectively. Children were encouraged by their schools to join the Gould League and to sign a pledge: 'I hereby promise that I will protect native birds and will not destroy their eggs. I also promise that I will endeavour to prevent others from injuring native birds and destroying their eggs'. Over a million Australians have been members of the Gould League since its inception.

To celebrate the centenary of the Gould League in 2009, former members were encouraged to sign up again and reaffirm their pledge.[10] By this time, two changes had been made to the league's identity: its emphasis has been widened to encompass broader environmental concerns and it now honours not only the work of John but also of Elizabeth and her contribution to the Gould legacy. Nearly 170 years ago, the extraordinary couple achieved more than any other naturalists of their time in recording the birds and mammals of Australia.

ENDNOTES

1. Gordon C. Sauer, *John Gould, the Bird Man: A Chronology and Bibliography*, Henry Sotheran Limited, London, 1982: xv.
2. Allan McEvey, 'John Gould's Contribution to British Art: A Note on its Authenticity', *Art Monograph 2*, Sydney University Press for the Australian Academy of the Humanities, 1973: 4.
3. http://australianmuseum.net.au/Gould-and-his-contribution-to-science, viewed 30 April 2010.
4. Libby Robin, Robert Heinsohn and Leo Joseph (eds), *Boom and Bust: Bird Stories for a Dry Country*, CSIRO Publishing, Collingwood, Victoria, 2009: 10.
5. John Gould, *Handbook to the Birds of Australia*, London, 1865, Preface: xxiv.
6. John Gould, *The Mammals of Australia*, vol. 1, London, 1845.
7. Robin, Heinsohn and Joseph (eds), *op. cit.*: 3.
8. Penny Olsen, *Glimpses of Paradise: The Quest for the Beautiful Parakeet*, National Library of Australia, Canberra, 2007.
9. *The Mammals of Australia*, *op. cit.*
10. The Gould League Pledge, 1909, http://australianmuseum.net.au/The-Gould-League-of-Bird-Lovers, viewed 30 April 2010.

PORTFOLIO
AUSTRALIAN BIRDS

Aquila fucosa
(Wedge-tailed Eagle)
nla.aus-f4773-1-s132

Haliastur leucosternus
(White-breasted Sea-Eagle)
nla.aus-f4773-1-s138

Pandion leucocephalus
(White-headed Osprey)
nla.aus-f4773-1-s142

Lepidogenys subcristatus
(Crested Hawk)

nla.aus-f4773-1-s180

Strix personata
(Masked Barn Owl)
nla.aus-f4773-1-s188

Athene boobook
(Boobook Owl)
nla.aus-f4773-1-s194

Athene rufa
(Rufous Owl)
nla.aus-f4773-1-s202

Hirundo neoxena
(Welcome Swallow)
nla.aus-f4773-2-s32

Chelidon ariel
(Fairy Martin)
nla.aus-f4773-2-s36

Alcyone azurea
(Azure Kingfisher)
nla.aus-f4773-2-s57

Gymnorhina tibicen
(Piping Crow-shrike)
nla.aus-f4773-2-s100

Pachycephala glaucura (Grey-tailed Pachycephala)
nla.aus-f4773-2-s139

Falcunculus frontatus
(Frontal Strike-tit)
nla.aus-f4773-2-s167

Gerygone albogularis
(White-throated Gerygone)
nla.aus-f4773-2-s204

Petroica phoenicea
(Flame-breasted Robin)
nla.aus-f4773-3-s17

Erythrodryas rhodinogaster
(Pink-breasted Wood-Robin)
nla.aus-f4773-3-s7

Malurus longicaudus
(Long-tailed Wren)
nla.aus-f4773-3- s45

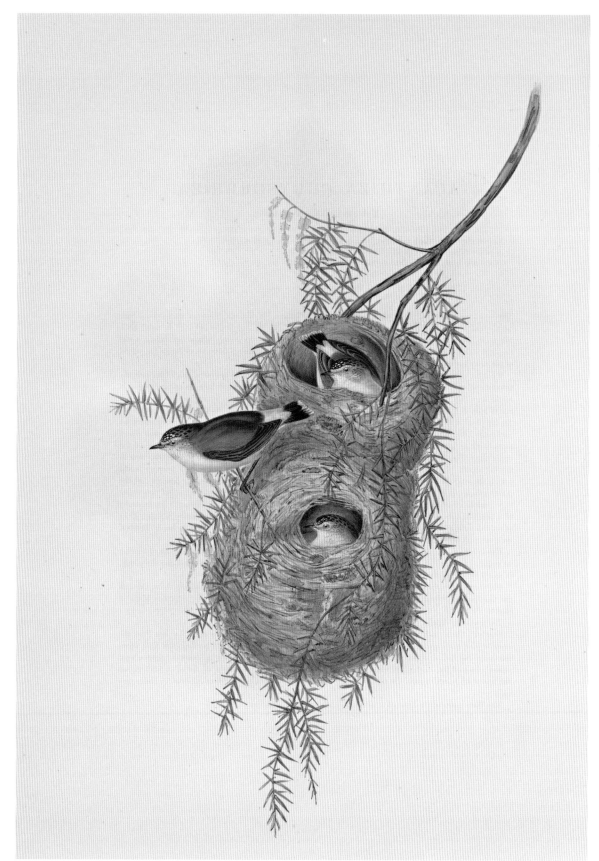

Acanthiza chrysorrhoea
(*Yellow-tailed Acanthiza*)
nla.aus-f4773-3-s133

Estrelda bichenovii
(Bicheno's Finch)
nla.aus-f4773-3-s167

Ptilonorhynchus holosericeus (Satin Bower Bird)
nla.aus-f4773-4-s28a

VOLUME IV
BIRDS

Neomorpha gouldii
(Gould's Neomorpha)
nla.aus-f4773-4-s48

Entomophila rufogularis
(Red-throated Honey-
eater)

nla.aus-f4773-4-s114

Anthochaera carunculata
(Wattled Honey-eater)
nla.aus-f4773-4-s120

Tropidorhynchus corniculatus (Friar Bird)
nla.aus-f4773-4-s126

Entomyza albipennis
(White-pinioned Honey-
eater)
nla.aus-f4773-4-s149

Cacatua galerita
(Crested Cockatoo)
nla.aus-f4773-5- s7

Cacatua leadbeateri
(Leadbeater's Cockatoo)
nla.aus-f4773-5- s9

Cacatua eos
(Rose-breasted Cockatoo)
nla.aus-f4773-5-s13

Licmetis nasicus
(Long-billed Cockatoo)
nla.aus-f4773-5-s15

Calyptorhynchus banksii
(Banksian Cockatoo)
nla.aus-f4773-5-s19

Platycercus semitorquatus
(Yellow-collared Parrakeet)
nla.aus-f4773-5-s43

Platycercus pennantii
(Pennant's Parrakeet)
nla.aus-f4773-5-s51

Platycercus exlmius
(Rose-hill Parrakeet)
nla.aus-f4773-5-s59

Platycercus pileatus
(Red-capped Parrakeet)
nla.aus-f4773-5-s70

*Nymphicus Novae-Hollandiae
(Cockatoo Parrakeet)*
nla.aus-f4773-5-s97

Pezoporus formosus
(Ground Parrakeet)
nla.aus-f4773-5-s99

Trichoglossus swainsonii
(Swainson's Lorikeet)
nla.aus-f4773-5-s104

Lopholaimus antarcticus
(Top-knot Pigeon)
nla.aus-f4773-5-s130

Peristera histrionica
(Harlequin Bronzewing)
nla.aus-f4773-5-s141

Talegalla lathami
(Wattled Talegalla)
nla.aus-f4773-5-s163

Megapodius tumulus
(Mound-raising Megapode)
nla.aus-f4773-5-s168

Synoicus australis
(Australian Partridge)
nla.aus-f4773-5-s190

*Dromaius
novae-hollandiae (Emu)*
nla.aus-f4773-6-s7

Haematopus longirostris
(White-breasted
Oyster-catcher)
nla.aus-f4773-6-s21

Lobivanellus lobatus
(Wattled Pewitt)
nla.aus-f4773-6-s25

Recurvirostra rubricollis
(Red-necked Avocet)

nla.aus-f4773-6-s61

Geronticus spinicollis
(Straw-necked Ibis)
nla.aus-f4773-6-s97

Grus australasianus
(Australian Crane)
nla.aus-f4773-6-s103

Platalea regia
(Royal Spoonbill)
nla.aus-f4773-6-s107

Porphyrio melanotus
(Black-backed Porphyrio)
nla.aus-f4773-6-s145

Parra gallinacea
(Gallinaceous Parra)
nla.aus-f4773-6-s157

Cygnus atratus
(Black Swan)
nla.aus-f4773-7-s17

Anas punctata
(Chestnut-breasted Duck)
nla.aus-f4773-7-s27

Biziura lobata
(Musk-duck)
nla.aus-f4773-7-s41

Diomedea exulans
(Wandering Albatros)
nla.aus-f4773-7-s82

Thalassidroma melanogaster (Black-bellied Storm Petrel)

nla.aus-f4773-7-s131

Pelicanus conspicillatus
(Australian Pelican)
nla.aus-f4773-7-s155

Spheniscus minor
(Little Penguin)
nla.aus-f4773-7-s175

Spheniscus undina
(Fairy Penguin)
nla.aus-f4773-7-s177

Strix candida
(Grass-owl)
nla.aus-f4773-8-s11

Malurus coronatus
(Crowned Wren)
nla.aus-f4773-8-s50

Ptilonorhynchus rawnsleyi (Rawnsley's Bower-bird)
nla.aus-f4773-8-s78

Casuarius bennetti
(Bennett's Cassowary)
nla.aus-f4773-8-s158

PORTFOLIO
AUSTRALIAN MAMMALS

Echidna hystrix
(Spiny Echidna)
nla.aus-vn760101-1-s54

Myrmecobius fasciatus
(Banded Myrmecobius)
nla.aus-vn760101-1-s59

VOLUME I
MAMMALS

Tarsipes rostratus
(Long-nosed Tarsipes)
nla.aus-vn760101-1-s61

Choeropus castanotis
(Chestnut-eared
Choeropus)
nla.aus-vn760101-1-s63

Peragalea lagotis
(Large-eared Peragalea)
nla.aus-vn760101-1-s65

Perameles fasciata
(Banded Perameles)

nla.aus-vn760101-1-s67

Phascolarctos cinereus
(Koala, head)
nla.aus-vn760101-1-s78

Phalangista fuliginosa
(Sooty Phalangista)
nla.aus-vn760101-1-s83

Phalangista vulpina
(Vulpine Phalangista)
nla.aus-vn760101-1-s85

Phalangista Cooki
(Cook's Phalangista)
nla.aus-vn760101-1-s90

Phalangista Viverrina
(Viverrine Phalangista)
nla.aus-vn760101-1-s92

Phalangista laniginosa
(Woolly Phalanger)
nla.aus-vn760101-1-s95

Cuscus brevicaudatus
(Short-tailed Cuscus)
nla.pic-vn760101-1-s97

Belideus flaviventer
(Long-tailed Belideus)
nla.aus-vn760101-1-s101

Belideus sciureus
(Squirrel-like Belideus)
nla.aus-vn760101-1-s103

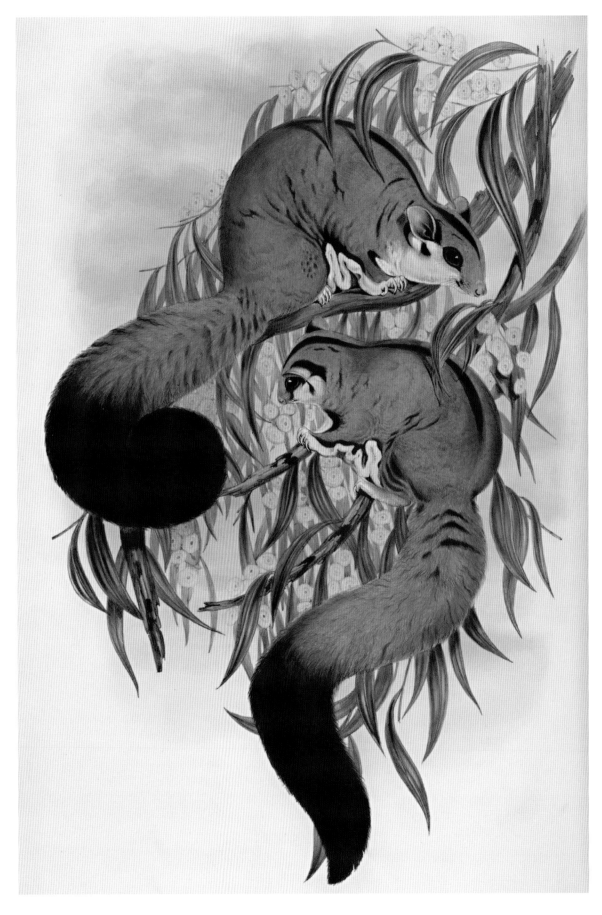

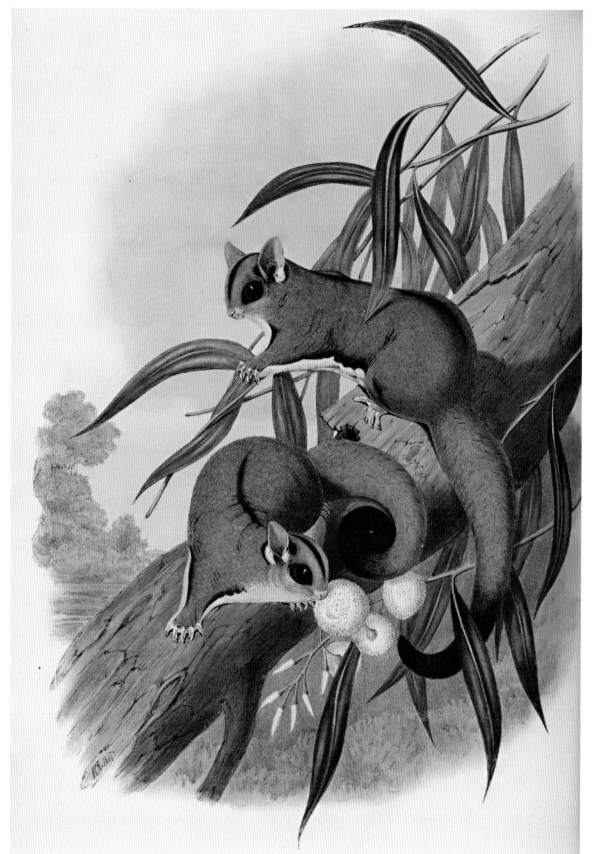

Belideus breviceps
(Short-haired Belideus)
nla.aus-vn760101-1-s105

Acrobates pygmaeus
(Pygmy Acrobates)
nla.aus-vn760101-1-s111

Dromicia gliriformis
(Thick-tailed Dromicia)
nla.aus-vn760101-1-s113

Dromicia concinna
(Beautiful Dromicia)
nla.aus-vn760101-1-s115

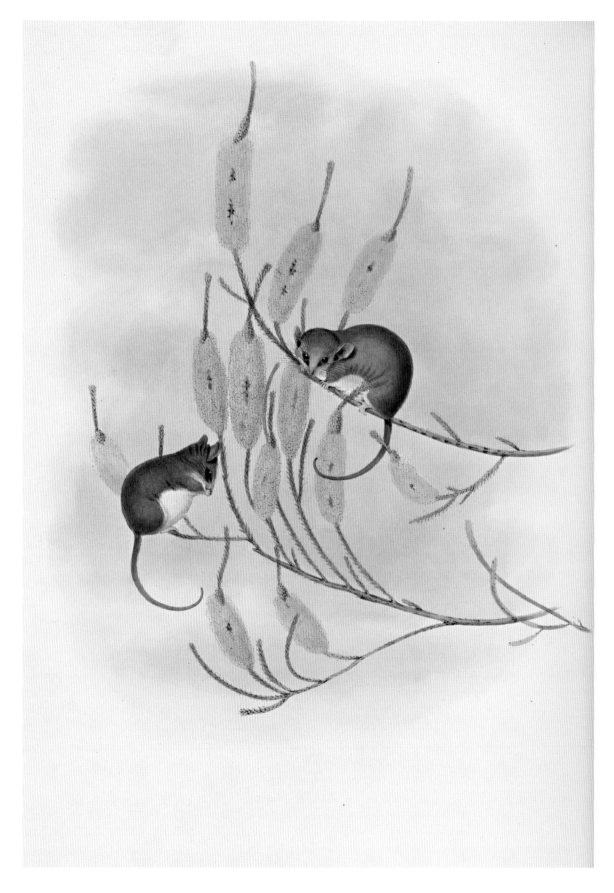

Phascogale penicillata
(Brush-tailed Phascogale)
nla.aus-vn760101-1-s117

Antechinus leucopus
(White-footed
Antechinus)

nla.aus-vn760101-1-s125

Antechinus unicolor
(Dusky Antechinus)
nla.aus-vn760101-1-s129

Antechinus apicalis
(Freckled Antechinus)
nla.aus-vn760101-1-s133

Antechinus flavipes
(Rusty-footed Antechinus)
nla.aus-vn760101-1-s135

Antechinus fuliginosus
(Sooty Antechinus)
nla.aus-vn760101-1-s137

Antechinus murinus
(Murine Antechinus)
nla.aus-vn760101-1-s141

Podabrus macrourus
(Large-tailed Podabrus)
nla.aus-vn760101-1-s147

Podabrus crassicaudatus
(Thick-tailed Podabrus)
nla.aus-vn760101-1-s149

Sarcophilus ursinus
(Ursine Sarcophilus)
nla.aus-vn760101-1-s151

Dasyurus maculates
(Spotted-tailed Dasyurus)
nla.aus-vn760101-1-s153

Dasyurus Viverrinus
(Variable Dasyurus)
nla.aus-vn760101-1-s155

Thylacinus cynocephalus
(Thylacine, complete)
nla.aus-vn760101-1-s163

Phascolomys wombat
(Wombat, head)
nla.aus-vn760101-1-s166

Phascolomys wombat
(Wombat, complete)
nla.aus-vn760101-1-s168

Phascolomys latifrons
(Broad-fronted Wombat,
head)

nla.aus-vn760101-1-s171

Phascolomys latifrons
(Broad-fronted Wombat,
complete)

nla.aus-vn760101-1-s173

Phascolomys lasiorhinus
(Hairy-nosed Wombat)
nla.aus-vn760101-1-s175

Macropus major
(Great Grey Kangaroo)
nla.aus-vn760101-2-s8

Macropus fuliginosus
(Sooty Kangaroo)
nla.aus-vn760101-2-s19

Osphranter rufus
(Great Red Kangaroo, pair)

nla.aus-vn760101-2-s23

Halmaturus ruficollis
(Rufous-necked Wallaby)

nla.aus-vn760101-2-s39

Halmaturus Bennetti
(Bennett's Wallaby)
nla.aus-vn760101-2-s45

Halmaturus Greyi
(Grey's Wallaby)
nla.aus-vn760101-2-s50

Halmaturus agilis
(Agile Wallaby)
nla.aus-vn760101-2-s62

Halmaturus Derbianus
(Derby's Wallaby)
nla.aus-vn760101-2-s70

Halmaturus Thetidis
(Pademelon Wallaby)
nla.aus-vn760101-2-s76

*Halmaturus Billardieri
(Tasmanian Wallaby)*
nla.aus-vn760101-2-s82

Petrogale penicillata
(Brush-tailed Rock
Wallaby, head)
nla.aus-vn760101-2-s90

Petrogale penicillata
(Brush-tailed Rock
Wallaby, complete)
nla.aus-vn760101-2-s92

Petrogale lateralis
(Stripe-sided Rock Wallaby)
nla.aus-vn760101-2-s95

VOLUME II
MAMMALS

Petrogale xanthopus
(Yellow-footed Rock Wallaby)
nla.aus-vn760101-2-s99

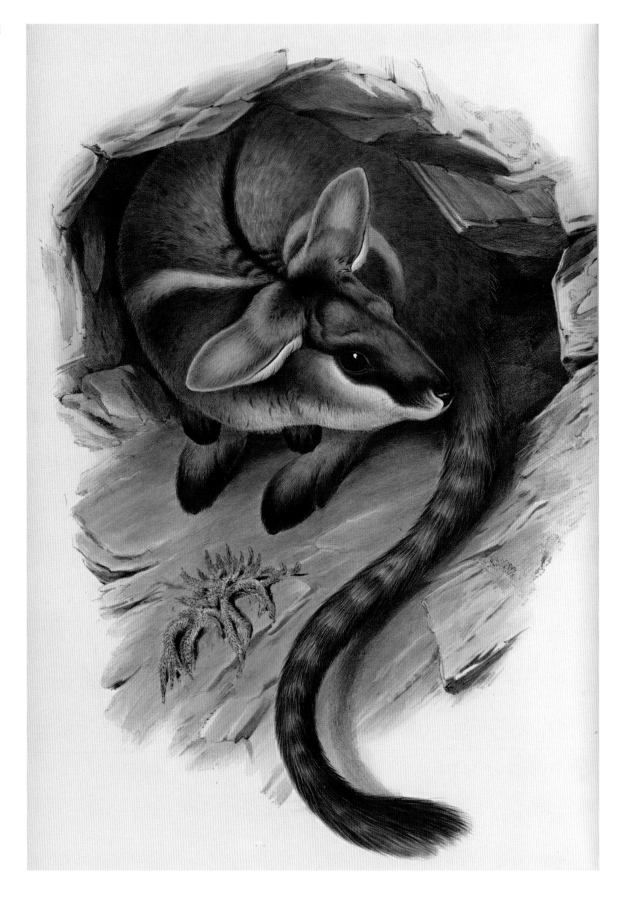

Petrogale inornata
(Unadorned Rock Wallaby, head)
nla.aus-vn760101-2-s103

Petrogale inornata
(Unadorned Rock Wallaby)
nla.aus-vn760101-2-s105

Dendrolagus ursinus
(Black Tree-Kangaroo)
nla.aus-vn760101-2-s111

VOLUME II
MAMMALS

Lagorchestes fasciatus
(Banded Hare-Kangaroo)
nla.aus-vn760101-2-s127

Lagorchestes conspicillata
(Spectacled Hare Kangaroo)
nla.aus-vn760101-2-s134

Bettongia penicillata
(Jerboa Kangaroo)
nla.aus-vn760101-2-s138

Hapalotis penicillata
(Pencil-tailed Hapalotis)
nla.aus-vn760101-3-s17

Hapalotis Mitchellii
(Mitchell's Hapalotis)
nla.aus-vn760101-3-s25

Mus fuscipes
(Dusky-footed Rat)
nla.aus-vn760101-3-s29

VOLUME III
MAMMALS

Mus cervinipes
(Buff-footed Rat)
nla.aus-vn760101-3-s35

Mus manicatus
(White-footed Rat)

nla.aus-vn760101-3-s39

Mus Gouldi
(White-footed Mouse)
nla.aus-vn760101-3-s46

Mus albocinereus
(Greyish-white Mouse)
nla.aus-vn760101-3-s50

Pteropus poliocephalus
(Grey-headed Vampire)
nla.aus-vn760101-3-s64

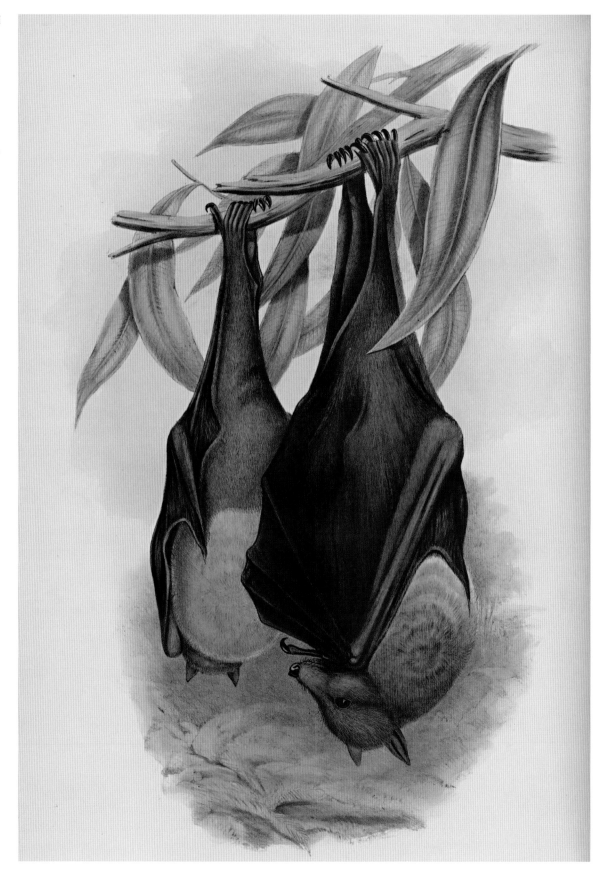

Pteropus conspicillatus
(Spectacled Vampire)
nla.aus-vn760101-3-s66

Scotophilus microdon
(Small-toothed Bat)
nla.aus-vn760101-3-s93

Arctocephalus lobatus
(Cowled Seal)

nla.aus-vn760101-3-s107

Stenorhynchus leptonyx
(Sea Leopard)

nla.aus-vn760101-3-s109

VOLUME III
MAMMALS

Canis dingo
(Dingo, head only)
nla.aus-vn760101-3-s111

Canis dingo
(Dingo, complete)
nla.aus-vn760101-3-s113

APPENDIX 1
GOULD'S MAJOR PUBLISHED WORKS[1]

A Century of Birds Hitherto Unfigured from the Himalaya Mountains (1830–1833)

The Birds of Europe (1832–1837)

A Monograph of the Ramphastidae, *or Family of Toucans* (1833–1835)

A Monograph of the Trogonidae, *or Family of Trogons* (1835–1838)

The Birds of Australia and Adjacent Islands (1837–1838)

A Synopsis of the Birds of Australia and the Adjacent Islands (1837–1838)

Icones avium, *or Figures and Descriptions of New and Interesting Species of Birds from Various Parts of the Globe* (1837–1838)

Charles Darwin, *The Zoology of the Voyage of HMS Beagle. Birds.* (1838–1841)

The Birds of Australia (1840–1848)

A Monograph of the Macropodidae, *or Family of Kangaroos* (1841–1842)

J. Gould's Monographie der *Ramphastiden* oder Tukanartigen Voegel (1841–1847)

Richard B. Hinds, *The Zoology of the Voyage of HMS Sulphur. Birds* (1843–1844)

A Monograph of the Odontophorinae, *or Partridges of America* (1844–1850)

Mammals of Australia (1845–1863)

A Monograph of the Trochilidae, *or Family of Humming-birds* (1849–1861)

The Birds of Asia (1849–1883)

The Birds of Australia. Supplement (1851–1869)

A Monograph of the Ramphastidae, *or Family of Toucans. Second edition* (1852–1854)

Supplement to the First Edition of a Monograph on the Ramphastidae, *or Family of Toucans* (1855)

A Monograph of Trogonidae, *or Family of Trogons. Second edition* (1858–1875)

An Introduction to the Trochilidae, *or Family of Hummingbirds* (1861)

The Birds of Great Britain (1862–1873)

An Introduction to the Mammals of Australia (1863)

Handbook to the Birds of Australia (1865)

An Introduction to the Birds of Australia (1873)

An Introduction to the Birds of Great Britain (1873)

The Birds of New Guinea (1875–1888)

Monograph of the Pittidae (1880–1881)

A Monograph of the Family of Trochilidae*, or Family of Humming-birds.* Supplement (1880–1887)

Introduction to Gould's Birds of Asia by Richard Bowdler Sharpe (1883)

ENDNOTES

1 This list is in chronological order. It has been adapted from an alphabetical listing in Gordon C. Sauer and Ann Datta (eds), *John Gould, the Bird Man: Correspondence, with a Chronology of his Life and Works*, Mansfield Centre, Connecticut, USA, in association with the Natural History Museum, London: vi–vii.

APPENDIX 2
GOULD'S AUSTRALIAN CHRONOLOGY 1838-1840

DATE	EVENT
1838	
16 May to 18 September	John and Elizabeth Gould and their party leave England for Australia aboard the *Parsee* and arrive in Hobart after a four-month journey from England
10 December	Gould leaves with Lady Franklin and Government Botanist Ronald Gunn aboard the government schooner bound for Macquarie Harbour; the party returns in two weeks due to bad weather. Elizabeth stays at Government House, Hobart, and paints Australian plants as backgrounds for ornithological illustrations
1839	
5 January	Gould and John Gilbert leave on a collecting trip to Launceston, Tamar River area and Bass Strait islands
4 February	Gilbert sent by Gould to collect in Western Australia; arrives aboard the *Comet* on 6 March
15 February	Gould leaves Hobart for Sydney aboard the *Potentate*
24 February to late March	Gould in Sydney; travels to Maitland for a week, then on to Stephen Coxen's property, Yarrundi, inland from Newcastle
4 April	Gould returns to Sydney via Maitland
11–23 April	Gould sails from Sydney aboard the *Susannah Anne* for Hobart; arrives 23 April
6 May	Franklin Tasman Gould born in Government House, Hobart
11 May	Gould leaves Hobart for Launceston; Elizabeth stays at Government House, Hobart

18 May	Gould boards the *Black Joke* for Adelaide
27 May	Gould arrives in Adelaide from Launceston
17 June	Gould leaves with Charles Sturt on expedition to Murray Scrubs (the Mallee)
23–26 June	Gould collects on ranges near upper Torrens
26 June	Gould travels towards Murray River
10 July	Gould back in Adelaide
19 July	Gould visits Kangaroo Island; then on to Hobart aboard the *Katherine Stewart Forbes*
20 August	Gould family leaves Hobart for Sydney aboard the *Mary Ann*
3 September	Gould family arrives in Sydney and stays with Dr George Bennett, Curator of the Australian Museum
14 September	Gould family leaves Sydney for Newcastle by steamer
15 September	Gould family arrives in Newcastle; stays there for a week and explores islands in Hunter River; Elizabeth Gould goes to Mosquito Island on 17 September with Gould
22 September	Goulds leave Newcastle for Maitland aboard the paddle steamer *William the Fourth*
23–28 September	Collecting and drawing around Maitland
29 September	Gould family leaves Maitland by cart for Yarrundi; stays overnight at Patrick's Plains, near Singleton
30 September	Gould family arrives at Yarrundi; spends two months there
November	Gould and party of six men, including two Aboriginal men, begin two-month expedition to the Liverpool Ranges and beyond; Elizabeth stays at Yarrundi to sketch specimens

1840

February to March	Gould family returns to Sydney; Gould visits Bong Bong (Moss Vale), Berrima and Camden
6 March	Australian Prospectus for *The Birds of Australia* issued in Sydney
9 April	Gould family and party leave Sydney aboard the *Kinnear*. Gilbert arrives back in Sydney on 3 May to find them gone
18 August	Gould family arrives back in London
1 December	First part of *The Birds of Australia* of 17 plates scheduled for publication in London

INDEX

Note: The plates in this publication are from John Gould's *The Birds of Australia*, volumes I–VIII and Supplement (1848), and *The Mammals of Australia*, volumes I–III (1863). The names of species are as they appear in Gould's publications. The plates can be viewed online through the National Library of Australia's website by typing **http://www.nla.gov.au/** and adding the identification number that appears next to each image.

Page numbers in bold refer to illustrations; scientific names and titles of publications and additional images are in italics.

A

Acanthiza chrysorrhoea **93**
Acrobates pygmaeus **158**
Adelaide 42
Alcyone azurea **85**
Amadina gouldiae **53**
Anas punctata **128**
Antechinus
    Dusky **163**
    Freckled **164**
    Murine **167**
    Rusty-footed **165**
    Sooty **166**
    White-footed **162**
Antechinus apicalis **164**
Antechinus flavipes **165**
Antechinus fuliginosus **166**
Antechinus leucopus **162**
Antechinus murinus **167**
Antechinus unicolor **163**
Anthochaera carunculata **98**
Aquila fucosa **76**
Arctocephalus lobatus **209**
Athene boobook **81**
Athene rufa **82**
Audubon, John James 8, 71
Australian Crane **123**

Australian Partridge **117**
Australian Pelican **132**
Azure Kingfisher **85**

B

Banded Myrmocobius **143**
Banded Perameles **147**
Banksian Cockatoo **105**
Beagle, HMS 5, 27, 29, 34
Belideus
    Long-tailed **155**
    Short-haired **157**
    Squirrel-like **156**
Belideus breviceps **157**
Belideus flaviventer **155**
Belideus sciureus **156**
Bennett, George **39**, 41, 45, 50, 51,
Bennett's Cassowary **138**
Bennett's Wallaby **183**
Benstead, James 31, 36, 37, 43, 51
betcherrygar 49
Bettongia penicillata **198**
Bicheno's Finch **94**
The Birds of Australia 50, 54, 56, 60, 64, 72
The Birds of Europe 21
The Birds of New Guinea 68
Biziura lobata **129**

Black-backed Porphyrio **125**
Black-bellied Storm Petrel **131**
Black Swan **127**
Black Tree-kangaroo **195**
Bowdler Sharpe, Richard 8, 14, 15, 22, 23, **64,** 65, 68
A Series of Gould's Most Attractive Works 71
Bower Bird
 Rawnsley's **137**
 Satin **95**
British Museum 9, 27, 29, 60, 64–5
Brush-tailed Phascogale **161**
Bubo maximus 22
Bucco grandis 17
budgerigar 5, 49, 54

C

Cacatua eos **103**
Cacatua galerita **101**
Cacatua leadbeateri **102**
Calyptorhynchus banksii **105**
Calyptorhynchus macrorhynchus **front cover**
Camarhynchus parvulus **29**
Canis dingo **211**, 212
Casuarius bennetti **138**
A Century of Birds from the Himalaya Mountains 14–19, 20, 21
Chelidon ariel **84**
Chestnut-eared Choeropus **145**
Choeropus castanotis **145**
cholera outbreak 66
Cinnyris gouldiae 18, **19**
Cockatoo
 Banksian **105**
 Crested **101**
 Great-billed Black **front cover**
 Leadbeater's **102**
 Long-billed **104**
 Rose-breasted **103**
Cook's Phalangista **151**
Cowled Seal **209**
Coxen, Charles 10, 29, 31, **41**, 49, 50
Coxen, Elizabeth *see* Gould, Elizabeth

Coxen, Stephen 10, 29, 31, 41, 45–6, 50
 suicide 58
Crested Hawk **79**
Cuscus brevicaudatus **154**
Cygnus atratus **127**

D

Darwin, Charles 5, **27**–30
 visit to Hobart 34
Dasyurus
 Spotted-tailed **171**
 Variable **172**
Dasyurus maculates **171**
Dasyurus Viverrinus **172**
Dendrolagus ursinus **195**
Dingo **211, 212**
Diomedea exulans **130**
Dromaius novae-hollandiae **118**
Dromicia
 Beautiful **160**
 Thick-tailed **159**
Dromicia concinna **160**
Dromicia gliriformis **159**
Drummond, Johnson 50, 58
Duck
 Chestnut-breasted **128**
 Musk-duck **129**

E

Echidna hystrix **40**, **142**
Emu **118**
Entomophila rufogularis **97**
Entomyza albipennis **100**
Erythrodryas rhodinogaster **91**
Estrelda bichenovii **94**
Euphema splendida **24**, **25**
extinction of species 71–3
Ewing, T.J. and Louisa 43

F

Fairy Martin **84**

Falcunculus frontatus **88**
finches 28–29
    Bicheno's **94**
    Gouldian **53**
Flame-breasted Robin **90**
Franklin, Sir John 5, **33**, 35, 42, 43
Franklin, Lady 5, **33**, 35, 37, 42, 43, 54–6, 68
Friar Bird **99**
Fringilla rodochroa **16**
Fringilla rodopepla **16**
Frontal Shrike-tit **88**

G

Gallinaceous Parra **126**
Geospiza scandens **28**
Geronticus spinicollis **122**
Gerygone albogularis **89**
Gilbert, John 13, 31
    correspondence with Gould 54, **55**
    death 57–8
    encounters with Aboriginal people 53, 54, 57–8
    instructions from Gould 39, 48, 50
    return to England 54
Gilbert's potoroo 59
Gilbert's Rat-Kangaroo **60**
Gould, Charles 31, 35, 62, 67, 68
Gould, Eliza 31, 33, 35, 65, **66**, 67
Gould, Elizabeth 11
    contributions to *A Century of Birds from the Himalaya Mountains* 15–19
    contributions to *The Zoology of the Voyage of H.M.S. Beagle* 29
    death 54
    diary **47**
    drawing skills 10, 11, 15, 50, 56
    finch named for 56
    marriage 11
    Mrs Elizabeth Gould's Collection of Drawings of Australian Plants, Flowers and Foliage **47**
    sun-bird named for 18
Gould, Franklin Tasman 5, 42, 43, 51, 54, 67
Gould, John **3**, 68
    childhood 7–8
    classification of Darwin's finches 28–9
    collecting activities 33, 37–9, 41–2, 49, 50
    death 68
    drawing skill, lack of 14–15
    encounters with Aboriginal people 38, 39, 48–50
    Fellowship of Royal Society 60
    Gould's Birds of Australia: A Fac-simile Reproduction in a Reduced Form **4**
    John Gould's House on Broad Street, London **65**
    journey to Australia 31–2
    marriage 11
    Moorhens with Young among Waterlily and Reeds **26**
    publications 13, 14–19, 21, 29, 60, 64, 68, **71**, 72, 214–5
    return to England 51, 54
    scientific papers 13
    shooting birds 8, 33, 38, 41
    significance of 5, 71–3
    taxidermy skills 8–10, 13, 26
Gould, John Henry 15, 31, 37, 43, 51, 66
Gould, Louisa 31, 33, 35, 42, **66**, 67
Gould, Sarah 54, **66**, 67
Gould League 5, 71, 73
Gouldian Finch **53**
Gould's Neomorpha **96**
Grey-tailed Pachycephala **87**
Grus australasianus **123**
Gymnorhina tibicen **86**

H

Haematopus longirostris **119**
Haliastur leucosternus **77**
Halmaturus agilis **185**
Halmaturus Bennetti **183**
Halmaturus Billardieri **188**
Halmaturus Derbianus **186**
Halmaturus Greyi **184**
Halmaturus ruficollis **182**
Halmaturus Thetidis **187**
Hapalotis Mitchellii **200**
Hapalotis penicillata **199**

Hare Kangaroo
    Spectacled **197**
    Banded **196**
Harlequin Bronzewing **114**
Hart, William 65
Helianthea osculans **63**
Hirundo neoxena **83**
Hobart 1, 32–5
    Government House, Hobarton **34**
    Vue de la rade de Hobart-Town, Ile Van-Diemende **32**
Honey-eater
    Red-throated **97**
    Wattled **98**
    White-pinioned **100**
hummingbird exhibition 62, **63**
Interior of the Humming-bird House, in the Gardens of the Zoological Society **63**
Hypsiprymnus gilberti **60**

I

Innes, Annabella 51

J

Jardine, Sir William **14**, 30, 65
Jemmy (Aboriginal companion) 48–50

K

Kangaroo
    Great Grey **179**
    Great Red **38**, **181**
    Jerboa **198**
    Sooty **180**
Koala **148**

L

Lagorchestes conspicillata **197**
Lagorchestes fasciatus **196**
Large-tailed Podabrus **168**
Large-tailed Trogon **23**
Lear, Edward **20**–3

Leichhardt, Ludwig 57, **58**
Blacks about to Attack Leichhardt's Camp, near the Gulf of Carpentaria, 1845 **59**
Lepidogenys subcristatus **79**
Licmetis nasicus **104**
Linnaean Society of London 2, 9, 13, 21
Lobivanellus lobatus **120**
Long-eared Peragalea **146**
Long-nosed Tarsipes **144**
Lopholaimus antarcticus **113**
Lophophorus impevanus **17**
Lyme Regis 1–2, 7
The Bay of Lyme Regis, Dorset **2**
lyrebirds 41, 48, 50, **51**

M

Macropus fuliginosus **180**
Macropus major **179**
Maitland 46, 48
Malurus coronatus **136**
Malurus longicaudus **92**
The Mammals of Australia 50, 59, 60, 64
Megapodius tumulus **116**
Melopsittacus undulatus 49, 54
Menura superba **51**
Mitchell's Hapalotis **200**
A Monograph of the Trochilidae, or Family of Hummingbirds 62
Mound-raising Megapode **116**
Mouse
    Greyish-white **205**
    White-footed **204**
Mus albocinereus **205**
Mus cervinipes **202**
Mus fuscipes **201**
Mus Gouldi **204**
Mus manicatus **203**
Muscicapa melanops **16**
Muscipeta princeps **16**
Musk-duck **129**
Myophonus horsfieldii **16**
Myrmecobius fasciatus **143**

N

Natty (Aboriginal companion) 48–50
natural selection 27–8
Nectarinia gouldiae 18
Neomorpha gouldii **96**
Newcastle 46
Nymphicus Novae-Hollandiae **110**

O

On the Origin of Species 27–9
Ornithorhynchus anatinus **40**
Osphranter rufus **38**, **181**
Owl
    Boobook **81**
    Eagle **22**
    Grass-owl **135**
    Masked Barn **80**
    Rufous **82**

P

Pachycephala glaucura **87**
Pandion leucocephalus **78**
Parra gallinacea **126**
Parrakeet
    Cockatoo Parrakeet **110**
    Ground Parrakeet **111**
    Pennant's Parrakeet **107**
    Red-capped Parrakeet **109**
    Rose-hill Parrakeet **108**
    Splendid Grass Parrakeet **25**
    warbling grass-parrakeet 49
    Yellow-collared Parrakeet **106**
Pelicanus conspicillatus **132**
Pencil-tailed Hapalotis **199**
Penguin
    Fairy **134**
    Little **133**
Peragalea lagotis **146**
Perameles fasciata **147**
Peristera histrionica **14**

Petrogale inornata **193**, **194**
Petrogale lateralis **191**
Petrogale penicillata **189**, **190**
Petrogale xanthopus **192**
Petroica phoenicea **90**
Pezoporus formosus **111**
Phalangista
    Cook's **151**
    Sooty **149**
    Viverrine **152**
    Vulpine **150**
Phalangista Cooki **151**
Phalangista fuliginosa **149**
Phalangista laniginosa **153**
Phalangista Viverrina **152**
Phalangista vulpina **150**
Phascogale penicillata **161**
Phascolarctos cinereus **148**
Phascolomys lasiorhinus **178**
Phascolomys latifrons **176**, **177**
Phascolomys wombat **174**, **175**
Picus shorii **17**
Piping Crow-shrike **86**
Platalea regia **124**
Platycercus exlmius **108**
Platycercus pennantii **107**
Platycercus pileatus **109**
Platycerus semitorquatus **106**
Podabrus crassicaudatus **169**
Podabrus macrourus **168**
Porphyrio melanotus **125**
Potorous gilberti 59
Prince, Edwin Charles 22–3, 26, 65, 67, 68
Pteropus conspicillatus **207**
Pteropus poliocephalus **206**
Ptilonorhynchus holosericeus **95**
Ptilonorhynchus rawnsleyi **137**
publishing process 21, 23, 26
Pygmy Acrobates **158**

R

Rat
    Buff-footed **202**
    Dusky-footed **201**
    White-footed **203**
Rawnsley's Bower-bird **137**
Recurvirostra rubricollis **121**
Red-necked Avocet **121**
Rhea darwinii **30**
Ripley Castle 9
Ripley Castle, Yorkshire **6**
Roper, John 57–9
Royal Society of London 2, 51, 60
Royal Spoonbill **124**

S

Sale of Australian birds 60, 64–5
Sarcophilus ursinus **170**
Scotophilus microdon **208**
Sea Leopard **210**
Short-tailed Cuscus **154**
Small-toothed Bat **208**
Spheniscus minor **133**
Spheniscus undina **134**
Spiny Echidna **142**
Stenorhynchus leptonyx **210**
Strange, Frederick 42, 50, 58
Straw-necked Ibis **122**
Strix candida **135**
Strix personata **80**
Sturt, Captain Charles 42, **43**
Swainson's Lorikeet **112**
Sydney 45–6, 51, 54
George Street, Sydney, Looking South **46**
Government House and Part of the Town of Sydney **45**
Synoicus australis **117**

T

Talegalla lathami **49**, **115**
Tarsipes rostratus **144**

Tasmanian tiger 72
taxidermy 3, 8–10
Thalassidroma melanogaster **131**
theft of bird specimens 60
Thick-tailed Podabrus **169**
Throsby, Charles 50
thylacine **72**, **173**
Thylacinus cynocephalus **72**, **173**
Top-knot Pigeon **113**
Tragopan satyrus **17**
Trichoglossus swainsonii **112**
Trogon macrouna **23**
Tropidorhynchus corniculatus **99**

U

Ursine Sarcophilus **170**

V

Vampire
    Grey-headed **206**
    Spectacled **207**
Vigors, Nicholas Aylward 9, 13–15, 18, 20

W

Wallaby
    Agile **185**
    Bennett's **183**
    Brush-tailed Rock **189**, **190**
    Derby's **186**
    Grey's **184**
    Pademelon **187**
    Rufous-necked **182**
    Stripe-sided Rock **191**
    Tasmanian **188**
    Unadorned Rock **193**, **194**
    Yellow-footed Rock **192**
Wandering Albatros **130**
Watson, Mary 31, 36, 43, 51, 56
Wattled Pewitt **120**
Wattled Talegalla **49**, **115**
Wedge-tailed Eagle **76**

Welcome Swallow **83**
White-breasted Sea-Eagle **77**
White-breasted Oyster-catcher **119**
White-headed Osprey **78**
White-throated Gerygone **89**
Wilson, Thomas B. 64–5
Windsor Castle 8–9
Wolf, Joseph 22, **67**
Wombat **174**, **175**
Broad-fronted **176, 177**
Hairy-nosed **178**
Woolly Phalanger **153**
Wren
    Crowned **136**
    Long-tailed **92**

Y

Yarrundi 29, 41, 43, 46, 48, 50,
    visit by United States Exploring Expedition 50
Yellow-tailed Acanthiza **93**

Z

Zoological Society of London 9, 10, 13, 14, 23, 27, 28
The Zoology of the Voyage of H.M.S. Beagle 29